OVERSIZE
STO
ACPL ITEM
DISCARDED
Y0-BXV-858
668.4
ENGINEERING PLASTICS
DO NOT REMOVE
CARDS FROM POCKET
ALLEN COUNTY PUBLIC LIBRARY
FORT WAYNE, INDIANA 46802
You may return this book to any agency, branch, or bookmobile of the Allen County Public Library.
DEMCO

ECE/CHEM/81

ECONOMIC COMMISSION FOR EUROPE
Geneva

ENGINEERING PLASTICS

UNITED NATIONS
New York, 1991

Allen County Public Library
900 Webster Street
PO Box 2270
Fort Wayne, IN 46801-2270

NOTE

The designations employed and the presentation of the material in this publication do not imply the expression of any opinion whatsoever on the part of the Secretariat of the United Nations concerning the legal status of any country, territory, city or area, or of its authorities, or concerning the delimitation of its frontiers or boundaries.

ECE/CHEM/81

UNITED NATIONS PUBLICATION

Sales No. E.91.II.E.32

ISBN 92-1-116520-2

CONTENTS

CONTENTS (continued)

LIST OF TABLES

ABBREVIATIONS USED FOR POLYMERS

ABS	acrylonitrile butadiene styrene copolymer
LCP	liquid crystal polymer
PA	polyamide or nylon
PBT	polybutylene terephthalate
PC	polycarbonate
PCTFE	polychlorotrifluorethylene
PEEK	polyetheretherketone
PEI	polyetherimide
PEK	polyetherketone
PES	polyethersulphone
PET	polyethylene terephthalate
PMMA	polymethyl methacrylate or acrylic
POM	polyacetal or polyoxymethylene
PPE	polyphenylene ether
PPO	polyphenylene oxide
PPS	polyphenylene sulphide
PS	polystyrene
PSU	polysulphone
PTFE	polytetrafluoroethylene
PU	polyurethane
PVC	polyvinyl chloride
PVdF	polyvinylidene fluoride
PVF	polyvinyl fluoride
RIM	reaction injection moulding
RRIM	reinforced reaction injection moulding
SAN	styrene acrylonitrile copolymer

PREFACE

The Chemical Industry Committee (now the Working Party on the Chemical Industry) of the Economic Commission for Europe, at its twenty-second session in October 1989, decided to undertake a study on Engineering Plastics (ECE/CHEM/74, para. 51).

The main purpose of the study is to analyse research and technical developments in the production and consumption of engineering plastics, such as polycarbonates, polyamides, polyesters, polyphenylenes, fluorine-containing plastics, acrylics and styrene copolymers in engineering industries, electronics, the automotive industry and construction.

The first ad hoc Meeting, held in January 1990, elaborated and agreed on the scope and outline of the study. It was decided that the study should cover the main groups of engineering plastics listed above, including market situations and their environmental impacts.

The second ad hoc Meeting, held in March 1991, considered the first draft of the study based on country contributions and made corrections to the text of the study as well as recommendations concerning its completion.

The present version of the study includes all material received from countries and considered by the two ad hoc Meetings.

The study consists of four main chapters. The first chapter is devoted to general properties, technologies of production and main areas of application of engineering plastics. The major group of plastics having similar chemical structures and physico-chemical properties were defined as: polycarbonates, polyamides and polyimides, polyacetates, polyesters, polyphenylene derivatives, fluorine-containing polymers acrylics, polyurethanes, and styrene copolymers.

Chapter II presents the existing technologies for the modification of engineering plastics in order to obtain the more specific properties needed for their applications. The different kinds of additive and blend technologies used for this purpose are described in this chapter.

The principle of "cradle-to-grave" is used in chapter III, where engineering plastics are considered from an environmental point of view. Information on toxicity, effects on the biosphere, and the flammability and disposal of engineering plastics are presented in this chapter.

Chapter IV describes the situation of engineering plastics in the markets of developed countries. All engineering plastics are considered, particularly acrylonitrile butadiene styrene copolymers and polyurethanes.

The Economic Commission for Europe expresses its thanks to Governments, international organizations, chemical companies and individual specialists who have provided valuable material for the study and given the secretariat the benefit of their advice.

Note: The mention of any firm or licensed process in the context of the present study does not imply endorsement by the United Nations.

CHAPTER I. GENERAL PROPERTIES, TECHNOLOGIES AND APPLICATION OF ENGINEERING PLASTICS

I.1 Introduction

Plastics are grouped together in this chapter, as the polymers within each group have chemical and structural features in common. The groupings are as follows:

POLYCARBONATES

POLYAMIDES AND POLYIMIDES
- Polyamides (including polyarylamides)
- Polyimides
- Polyetherimide

POLYACETALS

POLYESTERS
- Polybutylene terphthalate
- Polyethylene terephthalate
- Polyarylates
- Liquid crystal polymers

p-PHENYLENE GROUP DERIVATIVES
- Polyphenylene sulphide
- Polyphenylene oxide (including "Noryl") and polyphenylene ether
- Polysulphone
- Polyethersulphone and polyaryl sulphone
- Polyetheretherketone (with polyetherketone)

FLUORINE-CONTAINING POLYMERS
- Polytetrafluor ethylene
- Polychlorotrifluoroethylene
- Polyvinyl fluoride
- Polyvinylidene fluoride

ACRYLIC PLASTICS
- Polymethyl methacrylate

POLYURETHANES

STYRENE COPOLYMERS
- Styrene acrylonitrile copolymer
- Acrylonitrile butadiene styrene copolymer

The inclusion of "Noryl" and styrene copolymers is rather anomalous, as strictly speaking they should be treated as blends. However, they are included in this chapter since, in practice, they are treated as polymers in their own right.

I.2 Polycarbonates (PC)

Many PCs are chemically possible. The only commercially important ones are the bis-phenol A carbonates, prepared either by the exchange of ester dihydroxy compounds with diesters of carbonic acid and monofunctional aromatic or aliphatic hydroxy compounds, or by reaction of dihydroxy compounds with phosgene in the presence of acid acceptors.

PC is a transparent material with very high impact resistance, is resistant to creep, and shows rigidity and toughness up to 140° C. It has very good electrical insulation characteristics, is almost self-extinguishing, is physiologically inert, and has excellent hydrolytic stability.

Disadvantages include moderate flexural strength, low solvent resistance (particularly to paint solvents), high viscosity in processing (creating design and processing problems), and poor resistance to repeated impact. A component surviving a major impact may shatter if subjected to a second, much lower level, impact. Although transparent, it is actually pale yellow. A major disadvantage is its high cost.

Normal bis-phenol-A polycarbonates have a lower heat distortion temperature than some newer aromatic thermoplasts like polysulphones. Attempts to produce a PC of higher distortion temperature led to the production of a PC in which bis-phenol A was replaced by tetra bis-phenol A, but improvements in heat distortion temperature were obtained at the expense of impact strength and resistance to stress cracking. These PCs are used in headlamp reflectors and microwave dishes. PC copolymers involving bis-phenol S (thiodiphenol) have also been prepared because of their improved optical qualities.

Attack by petrol has limited full exploitation of PC car bumpers. This can be avoided by blending with PBT (see chapter on blends).

PC structural foams (expanded or cellular PCs) are also available, with low density and a high flexural strength to weight ratio. The products may be nailed and screwed like wood. Initial applications were intended for business machine housings, but there are many other applications, e.g. glass-reinforced grades have been used in water-ski shoes.

High impact resistance distinguishes PC from other transparent materials (glass, PS, acrylic). The extended temperature range, plus low density, make PC an important material for many applications.

PC can be foamed to give components with high impact strength, creep resistance and processability; reinforcements are usually glass fibres or particulate minerals. They are used in the structural elements of computers, telecommunications equipment, business machines and audio/video machines.

Transport: instrument panels, aircraft windows and canopies, lighting covers, car instrument back-lighting, automotive lenses, railway signals

Housewares:	blenders, coffeemaker cold water reservoirs, vacuum cleaner housings, serving implements, hair dryers, electric razors, microwave cookware, power tool housings, oven doors, front panels for domestic appliances
Glazing:	lighting, signs, inspection windows
Safety:	goggles, helmets, machine guards
Medical:	containers, tubing, dialysis components
Optical:	spectacle lenses, optical discs, lighting reflectors
Electrical/ electronic:	time switch, relay and battery covers, housings, connectors, coil formers, computer components, electrical power box back plates (in foil form) screen printing of membrane switch overlays
Telecoms:	telephone switching mechanisms, compact discs, business machine components, optical fibres
Security:	bus shelters, telephone booths, gym windows
Construction:	structurals sheet, strip-lighting covers, riot gear
Other:	compressed air lubricator bowls, laboratory animal breeding cages, milking pail covers, packaging for electronic components (antistatic grades)

Thin sheet PC is used for membrane switches, graphic overlay panels, keypads and protective eyewear. It may replace acrylic, PVC and CA altogether in these areas.

West European manufacturers include Bayer, Enimont, DSM and GE Plastics. Other manufacturers include Taijin Chemical Co., Mitsubishi Edogawa and Idemitsu Kasei (Japan), Dow (United States of America), and Policarbonatos do Brasil.

The global market for PC is estimated at 570,000 tonnes in 1990. Production and consumption more than doubled between 1980 and 1990, with a growth rate of about 11% per annum. This makes PC one of the very high growth areas among engineering plastics. Current capacity is 700,000 tonnes, but this is expected to be barely adequate by 1995. Major consumption regions are the United States (37%), western Europe (33%), and Japan (26%).

I.3 <u>Polyamides and Polyimides</u>

Polyamide (PA), or nylon, is best regarded not as a single plastic but as a family of plastics. All are based on CONH groups, but differ in composition according to the way in which they have been manufactured.

There is a very wide range of PA materials: fibres, crystalline plastics, amorphous plastics, adhesives, and rubbers. PA was first developed as a base for fibrous materials, and much PA output, particularly of PA 6 and PA 6/6, is still used for fibres. Only commercial formulations of engineering plastics are examined here.

The numbers defining different types of PA refer to the number of carbon atoms in the chemicals from which the polymers are made (e.g. 6, 6/6, 6/9, 6/10, 6/12, 11 and 12). Double numbers indicate two main constituent chemicals. Odd numbers of carbon atoms tend to inhibit crystallization. As a general rule, the higher the number, the higher the cost.

Processing conditions have a considerable influence on crystallinity, distinguishing PA from polymers like polyacetals and PCTFE. In PA, the crystallinity of a given formulation may vary by as much as 40%. The greater the crystallinity, the less the water absorption and the higher the abrasion resistance. The level of crystallinity also affects electrical and mechanical properties.

All PAs have very good abrasion resistance: this may be increased by the addition of external lubricants and by processing under conditions which develop a highly crystalline hard surface. All nylons are also hygroscopic, though the degree varies: PA 6 absorbs + 10% of its own weight of water at 100% relative humidity, PA 66 absorbs 8% and PA 11 absorbs 2%.

PAs have found steadily increasing applications for speciality purposes where toughness, rigidity, abrasion resistance, hydrocarbon resistance and reasonable heat resistance are important. PA-moving parts are self-lubricating, and silent running. Applications include gears, cams, bearings, bushes and valve seats. Nylon monofilaments are used in brush tufting, wigs, surgical sutures, sports equipment, braiding and outdoor upholstery. Films are used for packaging and salesroom displays. Extruded applications include cable sheathing, petrol lines in automotive products and as tension devices in drive belts.

The main PA types sold commercially are: PA 6, PA 6/6, PA 11, PA 12, and PA 4/6. It has been estimated that the breakdown of sales of various types of PA in western Europe in the late 1980s was as follows:

PA 6	46%
PA 6/6	42%
PA 11 and 12	9%
Other	3%

Source: Brydson, J.A., Plastics Materials, 5th ed., Butterworths, London, 1989.

The likely impact of DSMs new PA 4/6 plant is unknown. A breakdown of uses in western Europe in 1988 and 1989 was as follows:

	1988	1989	1000 tonnes
Extrusions			
Monofilaments	13	13	
Film	38	38	
Sheet, rod, tube	17	19	
Miscellaneous	6	6	
Injection moulding			
Domestic appliances	15	17	
Consumer goods	25	27	
Electrical & electronic	57	63	
Machinery parts	21	23	
Transport	79	86	
Miscellaneous	16	17	
Other	13	14	
Total PA	300	323	

Source: Modern Plastics International.

PA 6 is prepared from caprolactam by a ring-opening process. It is a crystalline material which is tough, resilient and creep resistant to dynamic loads. It has similar physical, chemical and electrical properties to PA 6/6 and, in the moulding shop is classed and handled in the same way, i.e. it is treated as a moisture-sensitive material. PA 6 has a lower melting point than PA 6/6 and a wider processing temperature range (25° C). It is slightly lighter in colour; has higher impact strength and slightly better low temperature properties; and is solvent, grease and dirt resistant. However, its resistance to dilute mineral acids is poorer. PA 6 is hygroscopic, which can be either advantageous or not, according to circumstances.

Many grades are available as the material may be extensively modified with particulate or fibrous fillers, plasticizers, etc. The use of additives enables flame retardant grades to be obtained. PA 6 is a reasonably good electrical insulator at room temperatures and in low humidity. PA 6 predominates in Germany. West European manufacturers include Akzo, BASF, EMS, Atocherm and BIP.

Applications include:

Automotive: gears, cams, hub caps, brake pedals, bearings, housings, car seat structural elements, handles

Mechanical: valves, impellers, blades, precision small mouldings

Furniture: castors, hinges

The Czech and Slovak Federal Republic was one of the first countries in the world to introduce the production of anionic poly-6-caprolactam by using its own process associated with the classical hydrolytic process. The CSFR researchers discovered the process for the anionic polymerization of 6-caprolactam which allows the two-component catalytic systems containing the initiator and the N-acyllactam-type activator component to be used, the latter being the actual growth centre of the polymerization. The process occurs as a sequence of rapid preacylative and neutralizing reactions.

Owing to the high rate of anionic polymerization, this process is best suited for use in both a discontinuous form for the production of polyamide mouldings and in a continuous form for the production of granules.

The new technology of production of polyamide mouldings of unlimited size and shape is based on the adiabatic process taking place in the whole range of temperatures below the polymer's melting point. The production of polyamide mouldings by this technique was introduced in the Czechoslovak firm Favorit Plazen in 1959-1960 and spread thereafter to other plants. The traditional initiator (6-caprolactam sodium salt, produced by the reaction of 6-caprolactam with sodium hydroxide, and the monomer being separated by distillation with water under a reduced pressure) was successively replaced by another initiator called Dilaktamát Spolana (80% solution of sodium dicaprolactamato-bis (2-methoxyethoxyethoxo) aluminate in toluene). In comparison with current alkaline initiators, the Dilaktamát has a number of technological advantages. It is liquid, so it is easy to handle, and it is not very sensitive to water and pollutants. Its use simplifies considerably the preparation of the reaction mixture which involves blending the constituents together. Dilactamát also allows to use polymer mouldings in non-chemical operations and has furthered the development of a complete line for the production of polyamide granules.

The technological process itself involves an initiator (which is dissolved in the melting process) and an activator (mono- or diisocyanate or other substances). After homogenization occurs, the reaction mixture is moulded and heated to between 150 and 180° C, depending on its size. Both the polymerization and crystallization will occur in the moulds. A relatively low polymerization temperature (below the melting point) of PA 6, as well as the postpolymerization taking place during the cooling are responsible for the low content of monomer and oligomers (below 4.0-4.5%).

The high rate of polymerization allows this technique also to be used for the rotary moulding of hollow objects such as bearings and tubes with a diameter greater than 1 m.

Polyamide produced by direct polymerization in moulds exhibits outstanding strength characteristics. The differences in the characteristics, compared with those of the hydrolytic polyamide, account for a higher molecular weight for a different structure, for higher crystallinity of the moulding PA 6 and for high tensile strength, hardness, and stiffness. Thermal hardness is improved in sliding properties and water sorption is significantly increased. A survey of some of the selected properties is given below.

Property	Dimension	Value (when dry)
density	g/cm^3	1.15 - 1.16
melting point	° C	216 - 225
water extractable content	% by weight	4.5 at the maximum
tensile strength	MPa	70 - 80
strain	%	10 - 20
notched toughness	$J.cm^{-2}$	0.2 - 0.3
impact strength	$J.cm^{-2}$	11 - 16

Source: Country contribution.

The values of the properties were determined according to the existing Czechoslovak standards (CSN) for PA 6. To a considerable extent, the properties depend on the kind of activators and, when multifunctional isocyanates are used, the dynamic mechanical properties may improve by as much as 50%. According to the buyer's requirements, the properties of PA 6 are modified by adding mineral fillers such as graphite powder, molybden sulphide, glass fibres, or oils. This is done to improve further the strength characteristics and sliding properties of PA 6.

Polymer mouldings are used in the production of shaped mouldings, such as driving and transmission gears, lining of cable wheels, auger feeders, sliding bearings of various types, pulleys, flanges, cams, parts of farm machines, and track bonds. A large proportion of the machine parts is still produced in the Czech and Slovak Federal Republic by cylindrical or plate machinery, generating much waste which implies further problems of waste processing.

The production of PA 6 in polymer mouldings in the Czech and Slovak Federal Republic has been continuously increasing. In 1985, the volume was 1,450 tonnes per year, and in 1989 it was 1,750 tonnes. The plants producing PA 6 are ZAZ Jaroměř (JARID), the Klecany Co-operative Farm (KLECAMID), PCHZ Žilina, and Škoda Plzeň (chiefly the filled types).

The production process of polyamide granules by continuous polymerization of 6-caprolactam was developed and introduced by the Spolana Neratovice Company. The polymer mixture is prepared with the use of Dilatamát Spolana as both initiator and activator of isocyanate type in a layout similar to that for polymer mouldings. Continuous polymerization then occurs in a screw reactor, and the extrusion of the polymerizate is cooled and granulated. The final stage involves postpolymerization at temperatures well below the PA 6 melting point. This new technological principle eliminates the extraction of monomer or vacuum demonomerization, which proves to be costly and harmful to the environment. The water extracted from the polymerizate is lower than 3%. The degree of polymerization can be regulated by the concentrations of the components of the catalytic system.

PA 6, in granulated form, is manufactured by the Spolana Neratovice Company under the name Spolamid, and is intended primarily for injection moulding. The standard type exhibits properties shown below.

Property	Dimension	Value when dry	Value (65% rel. moist.)
yield strength	MPa	78 - 80	33 - 39
tensile strength	MPa	45 - 55	55 - 60
elongation at break	%	30	100
Young's modulus (tension)	MPa	2 800	1 000
Young's modulus (flexural)	MPa	2 500	800 - 900
notched toughness	$J.cm^{-2}$	0.3	
impact strength	$J.cm^{-2}$	not broken	
melting point	° C	214 - 217	
relative thermal capacity	$J.kg^{-1}.K^{-1}$	$1.67 . 10^3$	

Source: Country contribution.

Other types produced include dyed, self-extinguishing and filled types. The properties of the standard and glass-filled Spolamid compare favourably with the respective type of Grilon, for example.

Spolamid is intended for mechanically-stressed parts in mechanical engineering, agriculture, transport, electronics, and other fields where injection moulding is used. The capacity of the production line is 1,400 tonnes annually and is fully exploited.

PA 6/6 is made by reacting hexamethylene diamine and adipic acid. It has the highest melting point of all PAs (except for PA 4/6 manufactured by DSM), the highest strength and stiffness, and the best stiffness retention with rising temperature. After moulding, it absorbs water (though less than PA 6). This causes swelling and an increase in toughness. It can be modified with glass fibres and with nucleating agents which give rapid crystallization. It is resistant to most solvents except strong acids or oxidizing agents. Outdoor exposure can cause colour fading and embrittlement unless the plastic is stabilized. Uses are similar to PA 6. The lower water absorption makes for better precision.

Glass reinforced super-tough PA 6/6 resins have been developed which are as tough as most amorphous plastics but lack their drawbacks: stress cracking, solvent sensitivity, low fatigue resistance and resistance to repeated impacts, and difficult processability. Uses include ski binding parts, blade supports for ice skates, fastener components, power tool housings, sprockets for motorcross bikes, roller skates, fan blades, and high impact-resistant gears.

West European makers include Akzo, BASF, Bayer, BIP, Du Pont, EMS, ICI, Jonylon, Monsanto, and Rhône-Poulenc.

PA 11 is based on castor oil. It performs well at high temperatures and has good UV resistance and insulation properties. Less hygroscopic than PA 6 or PA 6/6, it is less sensitive to changes in humidity. It has a fairly high heat distortion temperature and good low temperature impact resistance. It resists most solvents if they do not have extreme pH values.

Components may be used continuously at 65° C with some peaks at 100-130° C. These temperatures can be raised by stabilization with anti-oxidants. PA 11 has a lower melting point than PA 6 or PA 6/6. It also has very low moisture take-up, making it suitable for gear wheels for water wheels and for interior parts of water meters. It is used for powder spraying and fluidized bed coating. Monofilament uses include brush tufting, wigs, surgical sutures, sports equipment, braiding and outdoor upholstery. The only manufacturer is Atochem.

PA 12 is derived from butadiene. Its properties are broadly similar to PA 11, but PA 12 has a slightly lower melting point and density, and is slightly inferior to PA 11 in high temperature performance, UV resistance, heat distortion and low temperature impact resistance. Manufacturers include Atochem, EMS, and Huls.

PA 4/6 is made by reacting 1,4-diaminobutane with adipic acid. It has a high level of crystallinity and a high nucleation rate, giving higher impact strength than PAs 6 or 6/6. It has a relatively high tensile stress at break, a low degree of elongation at break, high specific gravity, and is very hard. DSM is the only manufacturer.

PA 6/10 is similar in properties to PA 11 but has greater tensile stress at yield. It is suitable for monofilaments, and is used in brush tufting, wigs, surgical sutures, braiding and outdoor upholstery because of its flexibility.

PA 6/12 is similar to PA 6, but with lower water absorption and melting point. The melt is more viscous and sets less rapidly on cooling, so thinner mouldings may be produced. It tends to be brittle. It is made by Du Pont and EMS.

"Glass-clear" PAs are copolymers of rather irregular structure having high performance and used for flow meter parts, filter bowls, pump casings, sanitary fittings, sight glasses, X-ray apparatus windows, gear wheels, milking machine covers, and water gauges for kettle jugs. Modified grades with improved resistance to alcoholic cleaning agents are used to make spectacle frames. They are made by Dynamit Nobel, Hoechst, and EMS.

The polyamides discussed so far are based on aliphatic chemicals. Polyarylamides (or aromatic polyamides) are based on aromatics and are prepared by reacting m-xylene with adipic acid. Most polyarylamide production is used for aramid fibres including "Kevlar", but some is used for mouldings.

They have the advantages of glossiness, impact and heat resistance, and stiffness. Water absorption is intermediate between PA 6 or 6/6 and PA 11 and 12. Thermal expansion is low, and mouldings have good dimensional stability and low warpage. The material has good electrical insulation properties.

However, they take longer to prepare than aliphatic based PAs, colour easily during polymerization, and tend to decompose before melting. Hydrolysis can occur when the material is exposed to steam. Chemical resistance characteristics are similar to PA 6/6. Uses include portable stereo cassette recorders, mowing machine components, electrical plugs and sockets, TV tuner blocks, and pulleys, shafts, and gears. Suppliers include Laporte and Solvay.

Polyimides (PI) are not dissimilar to polyamides in structure, but the functional group's branched structure leads to the production of polymers with high softening points and high thermal stability. They are made by condensing pyromellitic dianhydride with aromatic amines, particularly di-(4-amino-phenyl) ether.

PI is heat resistant, solvent resistant, flame resistant, and abrasion resistant, and it has excellent electrical qualities. It is claimed to be the highest temperature engineering polymer available, with continuous resistance up to 290° C and resistance to peak exposures up to 480° C.

Disadvantages are that it is difficult to mould, and that PI has limited resistance to hydrolysis, so that articles made from it may crack in boiling water or steam. (In such uses PEEK is the material preferred.) Because it is difficult to mould, it is usually used in sheet form. Uses include the following:

Automotive:	seal rings, thrust washers, seals, gaskets, valve seats, cam followers and piston rings, components for windscreen motors, and electric window lifts
Aerospace:	components for jet engines, e.g. compressor seals
Data processing equipment:	pressure discs, sleeves, bearings and guide rods friction elements
Mechanical:	valve shafts in cut-off valves, soldering and welding equipment
Heating and ventilating:	replacement of the reed valve system in light-weight poppet valves for freezer and air conditioning compressors
Other:	use as binder resin for grinding wheels.

Du Pont is the main west European supplier. Polyimide foams are produced by Monsanto, and PI fibres are produced by Rhône-Poulenc.

I.3.1 Polyetherimide

Polyetherimide is a representative of the group of aromatic polymers - the most heat-stable polymer materials. The valuable physico-mechanical and electrical properties of materials in this group are maintained over a temperature range of -196° to 260° C.

The sale of polyetherimides under the brand name Ultem was begun by the General Electric Plastics company in 1983. They are transparent amber-yellow polymers - products of the polycondensation of bisphenol A, 4,4-vinylenediamine and 3-nitrophthalic anhydride. The upper temperature limit for long-term utilization of items made from polyetherimides is 180° C, but they can withstand intermittant rises of temperature up to 200° C and short-term exposures to still higher temperatures.

Polyetherimides are amorphous engineering thermoplastics. They are characterized by excellent mechanical and electrical properties and are fire-resistant, self-quenching materials that give off only small amounts of smoke on burning. The oxygen index of polyetherimides is 47. An important property of these polymers is their high resistance to radiation. Polyetherimide film retains 94% of its tensile strength after irradiation with a dose of 400 Mrad. The polyetherimides are resistant to ultraviolet radiation.

The range of polyetherimides includes unfilled, filled and special composition materials. For filled materials use is made of carbon and glass fibres (up to 40%), polytetrafluoroethylene, graphite, calcium silicate, and aluminium powder. New brands of glass-fibre-reinforced (40%) polyetherimide, are marked by high tensile-strength values and high modules of elasticity in flexure and can be used in protracted service at 200° C.

The polyetherimides are processed on most types of standard equipment for injection moulding, extrusion and forming. They lend themselves to subsequent processing and can be joined to other materials with glues and by various types of welding. As a result of the good flow properties of the melt, items of complex shape can be produced by injection moulding.

The most important fields of application of aromatic polyimides are in the aerospace, electronic and electrical engineering industries. In the last few years the use of these polymers in the production of means of transport, in computer technology and in the making of office and medical equipment has increased. Items of printing equipment and flow-monitoring devices in hydraulic and pneumatic equipment are among special applications.

The volume of consumption of polyetherimides in 1985 (in thousand tonnes) was 0.3 in the United States, 0.2 in Western Europe and 0.5 in Japan. In general, the aromatic polyimides occupy third place in consumption among the group of highly thermostable engineering thermoplastics. Very high rates of growth in consumption by the mid-1990s are anticipated for the polyetherimides (21%), despite the fact that they are among the dearest polymers in this group ($US 31-33 kg).

I.4 Polyacetals

Polyacetal or Polyoxymethylene (POM)

POMs are produced by the polymerization of formaldehyde. They can be used well above 100° C although they have the lowest processing temperatures of all engineering plastics (190° C upwards). The base resin has good impact resistance at 23° C (though it is notch-sensitive), excellent fatigue qualities and is resistant to creep. It has a low coefficient of friction. POM mouldings are dimensionally stable, have good surface finish, and excellent machinability. Unfilled POM has a high relative density of 1.42 at 23° C - the highest of all engineering plastics. Acetal homopolymer resins show outstanding resistance to organic solvents, no effective solvent having been found for temperatures below 70° C.

POM is only moderately hygroscopic, but it should not be used for long periods in either hot water or in a steamy atmosphere. Raw material must be well dried before processing to avoid hydrolysis. POMs lack the impact toughness and abrasion resistance of polyamides. Electrical resistance is fair but not outstanding. Resistance to inorganic reagents is poor, and they should not be used in strong acids, strong alkalis, or oxidizing agents.

Different grades include toughened, improved hot water, friction, UV and weather resistant, and lubricated variants. Hoechst is marketing an electrically conductive copolymer.

Applications are usually found in structural, load-bearing uses and, where wear resistant and/or good surface finish is required. They also have a good record in uses involving hot air or hot water. They often reflect the early marketing objective of direct metal replacement, and include:

Plumbing applications:	pipe fittings, taps, snap-fit couplings
Automotive:	carburettors, fuel surge pots, window winders
Mechanical:	gears, cams, rollers, water and fuel pump housings and components, conveyor belt links, moulded sprockets and chains, blower wheels, fan blades
Household:	blow-moulded aerosol containers
Sports:	ski bindings
Retailing:	anti-theft tags
Other:	disposable lighters

Major European suppliers include Du Pont, BASF, and Hoechst/Celanese.

Global production of acetal polymers has been estimated at about 263,000 tonnes per annum in 1986, with capacity in the late 1980s at 340,000 tonnes, shared equally between Europe, North America and Japan. POM is one of the fastest growing engineering polymers, and forecasts indicate an increase in west European use of 31% (from 85,000 tonnes per annum in 1988 to 112,000 tonnes per annum in 1993).

I.5 <u>Polyesters</u>

The thermoplastic polyesters - polyethyleneterephthalate and polybutyleneterephthalate - are polyesters of ethylene glycol or butylene glycol and terephthalic acid. They are very similar in chemical structure and physico-mechanical properties. These materials are characterized by high impact strength, excellent hardness, flexibility, dimensional stability, anti-friction properties and chemical stability, together with a low degree of water absorption and cold flow. Both polymers are good dielectrics. Polyethyleneterephthalate is marked by a higher glass transition temperature and heat distortion point and a lower rate of crystallization, as a result of which it is more difficult to work than polybutyleneterephthalate. Casting grades of polyethyleneterephthalate - compounds of resin with additives that initiate and accelerate crystallization - were produced by the firm Akzo Plastics (Netherlands) in 1966. Polybutyleneterephthalate was first produced by the Celanese company in 1970.

Thermoplastic polyesters are processed by injection moulding, blow moulding, direct forming, stamping and deep drawing. The materials lend themselves well to mechanical pressing and surface finishing by fabric embossing and impression.

The bulk of the polyethyleneterephthalate produced in market-economy countries is used for making fibre. The percentage of total production thus used are 67% in the United States, 75% in western Europe and 63% in Japan. The remainder of the resin is used to produce various moulded products and the proportion of engineering grades among them is insignificant (1.3% in the United States, 1.8% in western Europe and 4.1% in Japan). The largest companies in the market-economy countries produce polyethyleneterephthalate for various purposes and are setting up plants to make various types of the resin.

The consumption of moulding grades of polyethyleneterephthalate and polybutyleneterephthalate in 1987 was estimated as follows (1,000 tonnes):

Product and field of application	United States	Western Europe	Japan
Polyethyleneterephthalate, total	630.5	302.0	232.5
Moulding products	8.5	5.5	9.5
Bottles	325.0	135.0	58.5
Trays	11.0	4.5	<1.0
Film	265.0	145.0	160.0
Other extruded products	21.0	12.0	5.0
Polybutyleneterephthalate for moulding*	34.0	26.5	32.5
TOTAL	664.5	328.5	265.0

* Excluding the resin used in producing mixtures and alloys.

According to data provided by specialists, the consumption of moulding grades of thermoplastic polyesters in market-economy countries is increasing at a higher rate than that of all the other high-tonnage engineering thermoplastics. This tendency will persist, according to the forecasts, up to the mid-1990s. The forecast consumption of polyethyleneterephthalate in the United States for moulding various products is shown below in thousand tonnes:

Field of application	1987	1989	1991	1993	Mean annual increase in the years 1988-1993, %
Bottles	325	465	544	626	11.5
Film	265	320	356	404	7.3
Trays	11	32	54	72	35.0
Other items	21	23	25	25	8.0
TOTAL	622	840	979	1 127	10.4

Consumption of moulding grades of thermoplastic polyesters in west European countries is as follows, in thousand tonnes:

Product and field of application	1980	1986	1990 (forecast)	Mean annual growth in production, %
Polyethyleneterephthalate, total	107	271	384	13.6
Film	90	140	185	7.5
Bottles	8	110	160	35.0
Moulded items	9	21	39	15.8
Polybutyleneterephthalate for moulding	7	22	23	12.6
TOTAL	114	293	407	13.6

Sources: Plastic World, 1987, 45, N.9, p. 22, 23.
Modern Plastics International, 1988, 18, N.1, p. 20-34.
Kunstoffe, 1987, 77, N.10, S.1004-1009.
Kunstoffe, 1986, 76, N.10, S.863-865.

In Japan it is expected that the use of thermoplastic polyesters for moulding will increase on the average by 7.3% per annum to reach about 305,000 tonnes by 1990.

Consumption of thermoplastic polyesters in the making of industrial products is small, amounting in 1987 to 75,000 tonnes in the United States and 25,000 tonnes in western Europe. However, according to forecasts by specialists, a considerable increase is to be anticipated in their use for these purposes - it will double in the United States by 1992 to about 150,000 tonnes (mean annual rate of increase from 1988-1992 about 15%) and rise in western Europe to 73,000 tonnes by 1995, with a mean annual rate of increase in the period 1988-1995 of about 24%.

Various grades of thermoplastic polyester have been designed for use as engineering materials, including glass-fibre-reinforced (5-55%) polyesters, carbonate and aramide fibres, glass-fibre mats, and modified elastomers with increased impact strength, which are also produced with glass-fibre and mineral fillers. The General Electric Plastics company is producing polycyclohexylenedimethyleneterephthalate, developed by the Eastman Chemical Plastics firm, in unfilled, reinforced and modified grades. Efforts are being made to develop compositions of thermoplastic polyesters with improved viscosity (particularly at low temperatures), high mechanical strength, increased heat stability, ageing resistance and yield strength and better processability. Coloured polymers are also being produced.

The main fields of application of engineering grades of polyethylene and polybutyleneterephthalates are transport and the electronic and electrical engineering industries. The structure of consumption in the United States and western Europe is shown below:

Field of application	Percentage of total consumption	
	United States	Western Europe
Transport	30	20
Electrical engineering and electronics	26	40
Industrial machine tools	20	*
Domestic electrical appliances	5	20
Consumer and leisure goods	5	*
Other items	14	20
TOTAL	100	100

Sources: Modern Plastics International, 1988, 18, N.1, p. 20-34.
Kunstoffe, 1986, 76, N.10, S.863-865.

* Included in "Other items".

According to forecasts by specialists, the volume of consumption of polyethylene and polybutyleneterephthalates in the automotive industry in the United States will increase by 93.6% by 1997.

In Japan, 65% of polyethyleneterephthalate and 60% of polybutyleneterephthalate is used in electrical engineering and electronics, 10% of polyethyleneterephthalate and 30% of polybutyleneterephthalate in the automotive industry, 9% of polyethylene terephthalate in the production of industrial equipment and 5% of polyethyleneterephthalate for consumer goods out of the total volume of consumption of each of these materials.

I.5.1 Polybutyleneterephthalate (PBT)

PBT is produced by reacting terephthalic acid with butane-1.4-diol. Because of the longer sequence of methyl groups in the repeating unit, the chains are both more flexible and less polar than PET.

PBT's properties differ from those of PET, with which it is often compared. PBT crystallizes fast, enabling fast cycling in injection moulds to produce stable crystalline products. Natural and reinforced PBT resembles Nylon 6/6 without the water uptake problems. The maximum temperature for continuous use is about 140° C.

PBT has good heat stability; good mechanical properties if reinforced; low water absorption; good electrical properties; good surface finish; good fatigue resistance; and very good solvent resistance.

However, PBT is brittle on impact at 23° C and can hydrolyze at melt temperatures (so material drying is vital). Fumes from processing operations must be extracted efficiently from the workplace.

About 90% of PBT is injection moulded. Being dimensionally stable, particularly in water, and with excellent resistance to oil, grease and petrol and good heat stability, PBT is appropriate for lubricated parts and under-the-bonnet applications. Good tracking resistance leads to electrical applications. Uses include:

Automotive:	fuel and ignition system parts, motor and mirror housings, door lock components, wiper arms, grilles, track modules
Mechanical:	pump housings, impellers, bearing bushings, measuring equipment gears
Electrical:	industrial plugs and sockets, motor insulation, pump, switch and circuit breaker housings, coil formers, safety plugs
Telecommunications:	telephones
Household:	power tools, housings for irons and kitchen appliances, self-cleaning ovens, stereo headphones
Orthopaedic:	lightweight wheelchairs
Furniture:	chair frames

Personal: cosmetic compacts

Sports: mid-soles for running shoes.

Suppliers include Atochem, Akzo, BASF, Bayer, Ciba-Geigy, Du Pont, EMS, Enimont, GE Plastics, Hoechst, Huls, and Rhône-Poulenc.

I.5.2 Polyethyleneterephthalate (PET)

PET can be obtained in either an amorphous or a crystalline state. Crystallization rates at sub-melt temperatures are rather slow compared with most other plastics, making the material easy to quench to a transparent form with very low crystallinity. This property however, makes it hard to mould fully crystalline parts at economic rates. PET is slightly hygroscopic.

PET is readily extruded to film or fibre form in highly oriented states. Orientation promotes crystallization, but the crystalline regions are smaller than the wave length of visible light. This means that although a film may be 25-30% crystalline, with the enhanced physico-mechanical properties that result, the product remains wholly transparent. The same applies to bottles.

Injection moulding PET grades are nucleated to improve crystallization rates. Ninety per cent of PET mouldings are glass fibre reinforced to overcome the low HDT ratings of the natural resin. Mica filling is also used to reduce distortion. The resulting products have low water uptake, good solvent resistance, and good electrical properties. However, rapid cooling of thin sections may result in incomplete crystallization and, hence, variable properties and possible dimensional changes. Flame retardant, reinforced and impact resistant grades are available.

PET mouldings are usually used in reinforced form to ensure good mechanical properties in temperatures above 140° C. Applications take account of PET's good electrical properties and solvent resistance, and include:

Automotive: light housings and ignition parts, car heater housings

Household: housings and components for toasters, coffee machines

Electrical: coil bobbins, motor and transformer housings, connectors, industrial plugs and sockets

Mechanical: gears, pump components

Other: chair shells, oven handles, water meter housings.

West European suppliers include Akzo, BASF, Bayer, BIP, Eastman Chemicals, Hoechst, EMS, Rhône-Poulenc, and ICI.

A related product to PBT and PET is PCT (polycyclohexylene terephthalate), claimed to offer the highest heat resistance of all polymer products. It is used for vapour phase solderable connectors, and is produced by GE Plastics.

I.5.3 Polyarylates

Polyarylates are made by reacting iso- and terephthalic acids and bisphenol A in the ratio 1:1:2.

They are clear, amorphous polyesters with a degree of molecular adaptability similar to PA. They are easier to process than many other engineering plastics, and can be injection moulded, blow moulded, and extruded. Key features are clarity, surface hardness, high dimensional stability, high impact, temperature and weather resistance, good electrical properties, and inherent flame retardance. Polyarylates show exceptionally rapid recovery after deformation. They are usually reinforced with up to 40% glass fibre, which gives higher tensile strength, stiffness, and heat deflection temperatures. A decorative finish can be given by metallizing.

Chemical resistance is good, and they are capable of resisting petrol, methanol, antifreeze, salt solutions and other fluids found in an automobile context.

Polyarylates were first produced in Japan in 1973. They appeared on the markets of the United States and western Europe in 1978-1979. By 1986, the following companies were supplying polyarylates and composite materials based on them: Unitika (Japan, U-polymer), Hooker Chemical Corporation (United States, Durel), Union Carbide Corporation (United States, Ardel), Celanese Chemical Corporation (United States, Durel), Solvay (Belgium, Arylef) and Bayer AG (Germany). In 1988, after it had obtained the rights for the production and sales of the Ardel polyarylates from Union Carbide Corporation, they were joined by Amoco Chemical Corporation (United States). In the same year Du Pont (United States) announced the beginning of production and sale of its own polyarylates under the trade name of Arylon and of Bexloy M. The firm developed its own technology for producing the resin but has not disclosed details of either the process or of the monomers used.

The polyarylates produced by the various companies have similar physico-mechanical and chemical properties. Their long-service temperature is 160° C. Suppliers are making considerable efforts to develop new compositions, different mixtures and alloys based on polyarylates.

The polyarylates are used as substitutes for metal, glass and other traditional materials in various fields of application and compete with thermoreactive and thermoplastic general-purpose engineering resins, particularly in view of their transparency and high weather-resistance with polycarbonates.

The main users of polyarylates are the electrical engineering and electronics industries. In this field, polyarylates are already being used for producing the most varied products - from printed circuits to connectors and large components made by injection moulding.

Another promising market for polyarylates is the motor-vehicle industry, where intensive development is now going on. They are being used for making headlight covers, components of the external parts of cars (the frames for the side lights, door handles, etc.) and internal parts (knobs, levers, locks, etc.). The Bexloy M polyarylate alloy proposed by the Du Pont company was developed specially for the automotive industry. It is being used for making bumpers and external panels (doors, wings) of vehicles. Parts made from this alloy are characterized by excellent external finish and easy colourability.

The price of Arylon polyarylate (4.41 $US/kg) is lower than the prices of the polyarylates produced by the other firms (4.48-6.17 $US/kg) and is close to that of such high tonnage engineering thermoplastics as polycarbonate, polyamide and polybutyleneterephthalate. The Du Pont company is proposing to reduce the prices for these polymers still further and this will make them more competitive. The firm considers that sales of Arylon polyarylates will be worth $US 250 million by 1995 and that consumption of these resins in the market-economy countries by the year 2000 will amount to 70-90,000 tonnes per annum (some specialists consider this figure too high). In 1988, consumption was estimated at 1,800 tonnes.

The main uses of polyarylates are as follows:

Automotive: lighting accessories, including headlight reflectors and lenses

Aircraft: lighting accessories

Airport: lighting accessories

Electrical/electronic: various components including fuses, switch housings and edge connectors

Other: clips and snap fasteners.

Polyarylates are marketed by Amoco, Bayer, Du Pont and Hooker.

I.5.4 Liquid crystal polymers (LCP)

The latest products in the group of highly thermostable engineering thermoplastics are liquid-crystal or self-reinforcing polymers, the industrial production of which began in the mid-1980s. These polymers are characterized by a high degree of molecular order, which is achieved by their molecular orientation in the solution (lyotropic) or in the melt (thermotropic) and is maintained in the products - fibres, films or castings. The degree of molecular orientation in liquid crystal polymers (LCPs) varies but, as a rule, it is higher in the thermotropic than in the lyotropic polymers and this affects the physico-mechanical properties of the items made from them which, like both solutions and melts of LCPs, are characterized by the high degree of anisotropy of their properties.

LCPs are aromatic polyesters based on phenol and hydroquinone, p-hydrobenzoic acid, terephthalic acid or hydroxynaphthoic acid. Different manufacturers may differ in their use of specific monomers, so LCPs should be regarded as a family of products. Commercial polymers differ in processing and end-use performance characteristics. However, they all resemble each other in that their totally aromatic structure provides unusually rigid, straight polymer chains, and these readily align to form highly ordered structures in both melt and solid states.

LCPs have low-melt viscosity, and hence easy processing, low-mould shrinkage; high softening points; rigidity and impact strength; and good inherent flame retardance. They have excellent chemical resistance, since they are unaffected by organic solvents, by acids up to 90% concentration, and by bases up to 50%. They do not stress-crack when exposed under load to most organic compounds. Barrier properties and resistance to weatherings are very good. Continuous use temperatures range from 220° up to 240° C. LCPs have good electrical properties, and are being used more in electrical applications.

LCPs also possess anisotropic (direction dependent), mechanical, and thermal expansion properties because of the ordering of the molecules. Properties along the flow direction are generally superior to those across it. Anisotropy can be reduced by the addition of fillers.

LCPs are tough, with less sensitivity to sharp notches than conventional thermoplastics, so they tolerate sharp corners and moulded-in notches more than most engineering polymers. They can be moulded to tight tolerances and have very good dimensional stability and resistance to warping. LCPs generally have low coefficients of friction with themselves, with metals, and with other materials.

The first industrial LCPs were developed by Du Pont in the form of a lyotropic aromatic polyamide - phenyleneterephthalimide, obtained from phenylenediamine and terephthaloyl chloride. Since 1971, this firm has been producing fibres from this polymer by solution-spinning in sulphuric acid. These fibres, under the trade name of Kevlar, have a high degree of rigidity and tensile strength, which is due to the high degree of orientation of the molecular chains and the presence between them of hydrogen bonds produced by the special features in the polymer structure. A general disadvantage of all lyotropic LCPs is the impossibility of transforming them into items with a complex shape. Up to the present, despite intensive work in that sphere, they can only have been made into fibres and films.

The first thermotropic LCPs were completely aromatic polyesters. The first industrial thermotropic LCP was a polymer of p-hydroxybenzoic acid developed by the Carborundum Company in the United States in 1970. It came on the market under the trade mark Ekonol in 1972. It possesses excellent physico-chemical properties, thermostability, thermal conductivity and dielectric strength and is self-lubricating. Its melting temperature (about 500° C) is higher than the temperature at which it begins to decompose (about 400° C). Ekonol can be processed by pressing or can be applied as coating by plasma spraying in a mixture with various metals - aluminium, bronze, nickel, etc. The protracted service temperature of items made from this polyester is 315° C and for a short time they can withstand up to 420° C.

The homopolyester Ekonol is processed poorly since its polymer chain has too rigid a structure. The next stage in the development of LCPs was to obtain copolymer polyesters based on p-hydroxybenzoic acid. Such products have been developed by the firms Dartco Manufacturing Inc. and Sumitomo Kagaku (Japan) which obtained licences for the production of the LCP Ekonol from the Carborundum Company. The products of both firms consist of copolymers of p-hydroxybenzoic acid, 4.4-dihydroxybiphenyl and terephthalic acid and can be processed by injection moulding.

The Dartco firm began to market its LCP at the end of 1984 under the trade name Xydar. The unfilled polymer melts at 421° C. The product marketed by Sumitomo Kagaku under the trade name of Ekonol has a melting temperature of 412° C and a lower heat distortion temperature than Xydar (293° C as against 355° C).

In 1985, the firm Celanese Speciality Operations (a branch of Celanese Corporation and since 1987 the Hoechst Celanese Corporation) began production of thermotropic LCPs under the trade name of Vectra. The firm does not disclose the composition of its products but it can be assumed that they are copolymers of 6-hydroxy-2-naphthoic acid and that one of them contains p-hydroxybenzoic links, while the second may be a polyamide ester. These polymers have a melting temperature of 250-285° C, which is lower than that of Xydar.

At the present time the American firm Amoco Chemical (which acquired the Dartco Company in 1987), Hoechst Celanese (trade mark Vectra), Du Pont (trade mark XL), Eastman Chemical and General Electric Plastics; the United Kingdom firm ICI; the Italian firm Granmont (trade mark Granlar); the German firms Bayer (trade mark KRU) and BASF (trade mark Ultrax), and the French firm Rhône-Poulenc, are producing highly heat-stable LCPs on an industrial and semi-industrial scale. The DSM firm in the Netherlands and Yunitika, Idemitsu Sekiyu Kagaku and Sumitomo Kagaku (all Japanese) LCPs at the development stage.

The highest tonnage products amongst LCPs are Xydar and Vectra, produced on an industrial scale since the end of 1984 and the beginning of 1986 respectively. They are characterized by excellent physico-chemical properties under load at room and high temperatures, creep resistance, very high moisture resistance and very high resistance to chemicals (they do not react with organic solvents, with acids in concentrations up to 90%, and with bases up to concentrations of 50%); they are fire-resistant even in the case of items with very thin walls and evolve little smoke.

In such cases as thermostability and fire-resistance, Xydar is better than Vectra. The heat distortion temperature of Xydar is 250-355° C (depending on the brand) and of Vectra 177-240° C, the protracted service temperatures are 240° and 200° C, respectively, and the oxygen indexes of the basic brands 42 and 35, respectively. Xydar can withstand temperatures of up to 315° C for a short time and its flexibility and impact strength are only reduced to an insignificant extent by long service at 220° C.

Vectra is cheaper than Xydar ($US 18-33/kg as against $US 40-62/kg) because of the lower cost of the naphthalene monomer and is also easier to work as a result of its lower melting temperature. It is possible to use LCPs as a basis for a large number of different compositions with tailormade properties (with carbon and glass fibre, talc and powder fillers), including items that cost less than the unfilled materials. The properties of LCPs change in accordance with the conditions and methods of processing since the degree of orientation of the molecules depends on this.

Hoechst Celanese, ICI and Amoco Chemical are studying the possibilities of using mixtures of LCPs with other thermoplastics, including polyethersulphones and polyarylketone, but the disadvantages of such mixtures are the poor compatibility of LCPs with other polymers. Amoco Chemical is also developing a liquid crystal fibre based on the Xydar polymer.

The Xydar LCP is being produced at a plant in Augusta, Georgia, with a capacity of 10,000 tonnes per annum and at a pilot plant with a capacity of 900 tonnes per annum in Neshanik Station, New Jersey, belonging to the Amoco Chemical Company. The Vectra material is being produced at a pilot plant with a capacity of 250,000 tonnes per annum by the Hoechst Celanese Corporation in the town of Summit, New Jersey. The construction of a plant for producing Vectra in Florence, Kentucky, has been announced. It will have an initial capacity of 2,300 tonnes per annum with a possibility of expansion up to 4,500 tonnes per annum.

Liquid-crystal, highly thermostable thermoplastics are new products on the plastics market and their fields of application are still extremely limited. Their properties would enable these materials to be used in various branches, particularly in electronics and aerospace, but so far only sample specimens have been developed for these branches. Up till now, the biggest field of application of LCPs has been for the production of Xydar utensils (saucepans, various containers) which can be moved straight from freezing compartments into ordinary or microwave ovens and then put on the table and afterwards washed in a dishwasher. This tableware, under the trade mark of Tupperware, is being produced in the United States by Premark Internationals Inc., which consumes 80% of the total amount of Xydar produced. Some items for military purposes are also being made from it.

The Vectra LCP is being used for making fittings for switchgear pillars working in an active medium and at high temperatures. The chemical industry is considered to be one of the promising fields for the use of liquid-crystal engineering thermoplastics, since by virtue of their high thermal stability and corrosion resistance and their low creep they can be used as a substitute for stainless steel and ceramics and also for fluoroplastics and thermoreactive resins, which are difficult to machine.

Electronics and electrical engineering are considered as among the main future fields of application of the highly thermostable LCPs. As a result of their high thermostability and dimensional stability over a wide range of temperatures and their ease of processing, which makes it possible to produce strong, thin-walled items in short moulding cycles, they can be used for making parts for electronic devices (connectors, integrated circuit boards, etc.), for encapsulating integrated circuits, etc. However, in the electronics industry the highly thermostable LCPs are encountering sharp competition from the cheaper epoxy resins, polyethersulphones and other engineering thermoplastics. For that reason it is expected that in the next few years they will be used only for manufacturing parts of electronic devices that operate at very high temperatures.

Other possible fields of application of highly thermostable and high-strength LCPs are the aircraft, space and war industries, fibre optics (optical-fibre cable sheaths, connectors, etc.), domestic appliances (parts of microwave ovens and compact disc players), sports goods (rackets and ski equipment, together with tackle for yachts), and film for barrier packaging. In the United States, prospects for using LCPs in the automotive industry are being investigated.

Care needs to be taken with reinforcement, but LCPs can be reinforced with glass and carbon fibres, and with particulate minerals. The main disadvantages of LCPs are poor abrasion resistance and very high price. Applications by end users can be summarized as:

Electrical:	connectors and encapsulants
Electronics:	mounted components required to withstand vapour phase reflow soldering, fibre optic sheathing, rigid data storage discs, moulded circuit boards
Automotive:	fuel, electronic and under-the-bonnet components
Aerospace:	radomes, fuel system parts, brackets, and electrical components
Housewares:	ovenable software, housewares and appliances
Process plant:	chemical plant components such as pulleys, bushes, bearings and wear blocks, packing materials in distillation columns.

At the present time, LCPs are considered to be the most promising of the new types of plastics, but growth in consumption is limited by their high cost and in some cases by the need to modify equipment for processing them, as well as by competition from other polymers and traditional materials. In 1984, on the eve of the appearance on the market of the first thermotropic liquid-crystal, highly thermostable thermoplastics, the Business Communications Co. in the United States estimated that the market for them in 1988 would be 4,500-9,000 tonnes (at $US 23-33/kg) but in fact consumption in 1988-1989 was estimated at about 450 tonnes per annum. According to an estimate by the American firm R.M. Kossoff and Associates, LCP consumption may reach 11-14,000 tonnes per annum provided that prices go down, otherwise it will be less than half that amount. A reduction in the prices of LCPs is the main prerequisite for expanding their consumption and may be achieved either by using cheaper chargestock, by manufacturing filled composites and alloys with other engineering thermoplastics or by improving the processing methods.

It should be borne in mind, however, that LCPs are at a very early stage of development, and that their use is likely to expand rapidly over the next few years. Some 25 companies are currently supplying or developing LCPs. The main ones in western Europe are BASF, Amoco Chemical, Hoechst and ICI.

I.6 p-Phenylene group derivatives

I.6.1 Polyphenylene sulphide (PPS)

Polyphenylenesulphides are crystalline aromatic polymers in which the rigid symmetrical chain consists of benzene rings linked in the para-position by atoms of sulphur. The first industrial method of obtaining polyphenylenesulphides was developed by Phillips Petroleum in the United States, which began their industrial production in 1973. The trade name is Ryton.

Commercial PPS is generally produced by the reaction of p-dichlorobenzene with sodium sulphide in a polar solvent. PPS is a "curable" thermoplast. It can be melt-processed and repeatedly melted if necessary (like other thermoplasts), but can also be heat-treated in air at high temperatures to form a possibly cross-linked thermoset-like material.

The polyphenylenesulphides belong to the highly heat-stable engineering thermoplastics. The protracted service temperature for items made from these materials is 200-240° C and they can withstand short-term temperatures up to 400° C.

Uncured PPS is very stiff but very brittle. Good mechanical properties are retained well above 200° C. It is resistant to solvents and burning, has good fatigue and abrasion resistance, and low friction. It also has good electrical insulation properties.

The polyphenylenesulphides combine excellent performance properties with good production and processing parameters. They are marked by high thermostability, extremely low creep, little shrinkage during moulding and insignificant water absorption. These polymers are not attacked by organic solvents or aqueous solutions of inorganic and organic acids and bases at temperatures below 200° C. They are fire-resistant (oxygen index 44), self-quenching materials and, when burning, evolve much less smoke than halogen-containing or other aromatic polymers.

Polyphenylsulphides are quite compatible with various fillers and the proportions of these added may be considerable quantities: the Bayer firm produces composite materials under the trade names Tedur KU-1-9511, containing 45% glass fibre, and KU-1-9521, containing 30% glass fibre, and 26% mineral filler.

After curing (entailing a change in colour from off-white to brown-black), PPS has improved solvent resistance and toughness, increased melt viscosity, and decreased crystallinity. The injection moulded form is usually reinforced with glass, carbon fibre or minerals. PPS is formable into amber/transparent films or 0.5 mm sheets.

Filled and unfilled polyphenylenesulphides are produced. The unfilled polymer is used mainly for coatings, which are characterized by their excellent hardness and resistance to corrosion and wear. Moulded polyphenylsulphide at room temperature is brittle and for that reason compositions with glass fibre filling are used when it is being made into various products. Polyphenylenesulphides at a temperature greater than 288° C possess high flow and an ability to wet the surface of glass fibre, thus increasing the strength of the composition. The introduction of various fillers modifies the colour, lustre and resistance to wear of the material. Compositions of polyphenylsulphide with glass fibre are used when good dielectric properties together with a high degree of mechanical strength are required.

One of the latest developments is the production of composite polyphenylsulphide materials, reinforced with mats made from long, randomly arranged carbon or glass fibres or spun unidirectional carbon and glass fibres (fabrics for moulding; wrapping cord). Not long ago, production began of broad braids made from polyphenylenesulphide resin, reinforced with oriented aramide fibres.

Polyphenylenesulphide is processed by injection or compression moulding. It can be re-worked without noticeable changes in its physico-mechanical properties. Moulded items are characterized by a high dimensional stability and a high-gloss finish.

The leading fields of application for polyphenylenesulphides are the electronics and electrical engineering industries and mechanical engineering, particularly for the chemical industry and transport. In 1987, the distribution of polyphenylenesulphides by field of application in Japan was as follows: electronics and electrical engineering (moulded items - switches, video parts, integrated circuits, electrical and heat insulation) - 40%; mechanical engineering (pump casings, valves, flow meters, parts of compressors and precision pumps, spray nozzles, ventilators for the chemical industry and cisterns) - 20%; automotive production (parts for carburettors, sensors, reflectors and lamp holders, bearings) - 35%; other items (parts of precision instruments - computers, optic and measuring instruments, cameras, tachometers, watches) - 5%. In the United States, the main fields of application of polyphenylenesulphides are also electrical engineering and electronics (about 38%). It is forecast that they will remain the most important for the next decade (in thousand tonnes).

Field of application	1988	1998
Electronics and electrical engineering	2.02	6.35
Machine building		
Industrial equipment	1.36	3.62
Means of transport	0.45	1.81
Chemical industry	0.45	1.36
Domestic electrical appliances	0.23	0.90
Others	0.90	4.54
TOTAL	5.41	18.58

Source: Revue de Plastiques Modernes 33, No. 9, p. 132-133.

Consumption of polyphenylenesulphides by main areas of application are as follows:

Automotive: exhaust gas return valves, carburettor parts, ignition plates, flow control valves for heating systems, fuel system components

Mechanical: aerospace uses, pump housings and parts, valve components, light reflectors

Electrical: coil formers, brush holders, switches, relays, electronic watch bases

Household: cooking appliances, hair dryer components

Medical: sterilizable medical, dental and laboratory equipment

Fibres: filters, flame-proof clothing and upholstery.

The thermosetting forms are used in compression moulding powders, as the binder in glass cloth laminates like the polymer base in heat resistant metal coatings.

The volume of consumption of polyphenylenesulphides in the market-economy countries in 1985 was 9.7 thousand tonnes. According to forecasts made by specialists, the mean annual rate of growth in the consumption of these materials by the mid-1990s will be 14.4%, and by 1995, the volume of consumption will exceed 37,000 tonnes. The price of polyphenylenesulphide with 40% glass fibre content is about $US 7/kg. Main west European producers are Phillips Petroleum, Hoechst, and GE Plastics. Bayer has announced electrically conductive grades.

I.6.2 Polyphenylene oxide (PPO), including "Noryl" and polyphenylene ether

Polyphenyleneoxide and polyphenyleneether are polyethers of 2,6-dimethylphenol and a copolymer of 2,6-dimethylphenol and 2,3,6-trimethylphenol. The materials are very similar in their chemical structure and their chemical, physical and mechanical properties are practically identical.

Commercial PPO is a misnomer. Polyphenylene oxide proper is virtually unprocessable and has no commercial importance. The designation PPO has become attached to poly-(2,6-dimethyl-p-phenylene oxide), which is an alkyl substituted polyphenylene oxide.

Commercial PPO is crystalline, with a low melting point. It is of comparatively low molecular weight and is essentially linear, though it contains some cross-linkages. It is attacked by halogenated and aromatic hydrocarbons. It is hydrolytically stable, and has exceptional dimensional stability. It has a low coefficient of thermal expansion, low moulding shrinkage and low water absorption, thus permitting moulding to close tolerances. High processing temperatures (280-330° C) are required.

Polyphenyleneoxide was first synthesized in 1956 and its industrial production was begun by the General Electric Plastics firm in the United States in 1964. Polyphenyleneether was developed and put on the market in 1979 by the Japanese firm Asahi Dow. Both thermoplastics have valuable service properties, although some disadvantages exist, one being the high degree of viscosity of the melt, as a result of which their processing by injection moulding is possible only at temperatures of 315-340° C, which is considerably higher than the alloy's softening point. These difficulties were overcome by developing modified grades of polyphenyleneoxide and polyphenyleneether by alloying them with other polymers, the first of which was polystyrene.

At the present time, composite materials combined with practically all polymers - polystyrenes, polyamides, polycarbonates, polyolefins, polysulphones, elastomers, polysiloxanes - are available. Modification makes processing easier (the temperature range is extended to 230-315° C), the cost of the material is reduced and its excellent physico-mechanical and dielectric properties are preserved, but its thermal stability and protracted operating temperature are reduced. In respect of a whole set of properties, the composite materials have proven to be preferable and have found wide industrial applications.

At the present time, over 170 different grades of modified polyphenyleneoxides and polyphenyleneethers are being produced. These are mixtures and alloys with various polymers which have a reduced melt viscosity, and improved impact strength, elasticity, chemical stability and processability (the improvement of this last property is achieved by introducing 1-50% of plastifiers). Fire-resistant grades, obtained by using fire retardants of various types (diaryloxy compounds, antimony trioxide and its mixtures with phosphates or phosphites, halogenated aromatic hydrocarbons) have heat stability at 90-150° C and impact strength of 200-300 J/m. Composites have been developed with mineral fillers and carbon and glass fibres. The introduction of 10-30% glass fibre increases the heat stability of modified polyphenyleneoxide and polyphenyleneether to 120-144° C, while the elasticity modules and tensile and flexure strengths are almost doubled. The grades of the polymers that are being produced are designed for various manufacturing processes, including the foaming of foamed plastics, which is brought about by introducing nitrogen or conventional foaming agents and also the obtaining of weather-resistant and coloured composites.

Modified polyphenyleneoxide and polyphenyleneether are processed by injection moulding, blown bubble extrusion, vacuum forming, and deep drawing. The materials lend themselves easily to mechanical processing (drilling, milling and sawing) and to coating techniques (lacquering, embossing), glueing (with various solvents and epoxy, silicone or urethane adhesives) and welding (ultrasonic, contact or electrical).

The consumption of modified polyphenyleneoxides and polyphenyleneethers in the market-economy countries in 1988 was 205,000 tonnes, of which about 45% was attributed to the United States. The structure of consumption of these materials in the United States and west European countries was as follows (%):

Field of application	Western Europe 1985	United States 1987
Transport	35.0	14
Electric engineering and electronics	25.0	15
Domestic electrical appliances	10.0	24
Office equipment	15.0	29
Industrial equipment	7.5	9
Other applications*	7.5	9
TOTAL	100.0	100

Sources: Kunststoffe, 1986, 76, No. 10, S.859-861.
Modern Plastics International, 1988, 18, No. 1, p. 20-34.

* Including consumer goods, goods used in the household and leisure goods.

In Japan the main spheres of application of polyphenyleneoxides and polyphenyleneethers (about 30%) are the automotive industry, automatic office equipment production and electric lighting fittings.

The price of PPO was too high to justify more than a very restricted application. This led to the introduction of the related "Noryl" PPO/PS alloy, which is generally used in the United States and western Europe. In recent years, the only sources of unmodified PPO have been the USSR (Aryloxa) and Poland (Biapen).

Noryl is an alloy of polystyrene and PPO. It should strictly be considered a blend, but it is treated here as the main (and in North America and western Europe the only) way in which PPO is used.

Unfilled Noryl has a good impact resistance, low density and low water uptake. It is inherently flame resistant. It has a high distortion temperature above the boiling point of water. It is an amorphous material which can be vacuum formed from extruded sheet. It can also be injection moulded and blow moulded. However, melt viscosities are high, inhibiting flow lengths in injection moulds and often requiring high processing temperatures.

Several glass- and mineral-filled grades are marketed. Other special grades are for extrusion and structural foam processes and for specific electrical and motor industry requirements.

Noryl's impact resistance, flame retardancy, and dimensional stability suit it for load-bearing uses in the automotive and appliance industries and for electrical and fluid handling applications. Uses include:

Automotive: dashboards, grilles, handles, air inlet and outlet grilles, outer mirror housings, steering column cladding, loudspeaker housings, nozzles, parcel shelves, radiator and expansion tanks in cooling systems, auto-electrical parts including cable connectors and bulb sockets

Electronic: business machines and computer housings

Electrical: timers, switches, conduits, busbars, connectors, switch cabinets, fuse boxes, housings for small motors, transformers and protective circuits

Radio/TV: coil formers, picture tube deflection yokes, insert card mountings

Fluid handling: components in pumps, meters, impellers, heat exchangers, filters

Other: washing machine parts (glass filled).

Noryl is marketed by G.E. Plastics. A similar material is marketed by Mitsubishi. PPO/styrene graft materials with similar properties but rather lower heat deflection temperatures are marketed by Asahi Glass. Production figures are not published, but production is thought to be somewhat over 200,000 tonnes per annum.

I.6.3 Polysulphone, polyethersulphone and polyarylsulphone (PSU)

PSUs are a group of materials based on p-phenylene groups, with varying degrees of spacing between the groups. They are prepared by polyetherification or polysulphonylation.

Polysulphones, polyethersulphones and polyarylsulphones constitute a group of engineering thermoplastics with a high degree of heat stability, known collectively as polysulphones. They are aromatic, heterocyclic polymers containing ether and sulpho groups in their main chain.

Industrial production of polysulphones was developed in the mid-1960s by the Union Carbide Corporation in the United States. At the present time, polysulphones, polyarylsulphones and composition materials are produced by Amoco Chemical in the United States, which acquired the production and sale rights from Union Carbide in 1986. Polyethersulphones are produced by the Minnesota Mining and Manufacturing Company and the Fluorocarbon Company in the United States, and ICI in the United Kingdom. In Japan, the Sumitomo Kagaku and Mitsui Toatsu Kagaku companies sell ICI's polyethersulphones.

All commercial PSUs are virtually amorphous, with high resistance to heat deformation. They have exceptional resistance to creep, good high temperature resistance, rigidity, and self-extinguishing properties. They are more heat resistant than PCs and have greater resistance to creep. They have, however, lower Izod strength and are much more expensive. Higher molecular weight grades show greater resistance to stress cracking, have better long-term strength under load and, in some cases, have better impact strength. Glass-fibre-filled PSUs have increased creep resistance and lower coefficients of thermal expansion.

The polysulphones have a high heat stability and resistance to oxidative breakdown and have good dielectric properties if maintained at temperatures up to 175° C and in a moist environment. They are resistant to aqueous solutions of acids and bases, synthetic detergents, oils and alcohols. They are self-quenching materials that evolve little smoke when burning. The polysulphones are distinguished by their high creep resistance and resistance to ionizing and high-frequency radiation. However, they are not resistant to ultraviolet radiation and, when used in the open air, have to have stabilizers incorporated in them.

PSUs are slightly hygroscopic. The high processing temperatures (320-415° C) required can lead to streaks and splash marks. Even at these temperatures, PSU melts are very viscous. Moulding can involve high frozen-in stresses, which can be remedied by annealing.

The temperature for the protracted use of polysulphone is 160° C (they can withstand up to 190° C for short periods). The temperature limit for polyethersulphones is 180° C for 20 minutes and about 200° C for 4-6 minutes; the limit for polyarylsulphones is 180° C.

Industry has now developed various grades of polymer in this group - transparent, non-transparent and coloured, unfilled and composition materials with glass fibre, mineral and PTFE fillers, and special sorts for electrodeposited coatings, for use in medicine, or with a high degree of resistance to creep. When they are reinforced with glass fibre (usually up to 20 or 30%) the elasticity modules of these polymers is greatly enhanced and their heat stability slightly improved. The heat distortion points of unfilled polysulphone and polysulphone containing 30% glass fibre are 174° C and 185° C, respectively. The corresponding figures for polyethersulphone are 203° C and 216° C.

Polysulphones are processed on standard equipment by injection moulding, extrusion, blow moulding and thermal moulding and pressing. When they are treated by injection moulding, 25-50% chopped waste can be introduced without affecting the processing parameters or changing the properties of the products. As a result of their low shrinkage coefficient on moulding, production of precision items from these materials may be carried out within fine limits.

PES combines high temperature performance with very stable mechanical properties up to 150° C, with HDT values around 200° C. Maximum temperature for continuous use is about 180° C.

It retains almost constant stiffness over a wide range of temperatures and time. This low creep behaviour benefits designers, as performance is readily predictable. PES is amorphous (and therefore transparent), with low shrinkage and distortion. It is affected by some organic solvents but not generally by inorganic chemicals.

Its main disadvantages are that properties can change somewhat with aging at higher temperatures and fatigue resistance is relatively poor. It is also one of the most expensive engineering plastics.

Polysulphone is used for making film by the solution casting method in dimethylformamide; polyarylsulphone is used to make coatings for solution spraying. These coatings are extremely hard and reistant to abrasion.

Items made from polysulphones may be subjected to such forms of subsequent processing as metallization, glueing, welding and coating.

The fields of application of polysulphones include electrical engineering, electronics, the automotive and aircraft industries, the production of industrial, medical and office equipment and the manufacture of domestic goods and packagings. About 40% of polysulphones are used in electronics and electrical engineering, assembly plates, moving parts in relays, parts of potentiometers, cable and condenser insulation, television sets, video and stereo apparatus, the antenna caps in radar sets and a large number of small items.

PSUs tend to be used in applications when requirements cannot be met by using PCs or aromatic polyethers. In many uses they replace ceramics, metals and thermosets rather than other thermoplasts. Applications include:

Electrical: coil bobbins, printed circuit boards, lamp bulb bases, TV components

Automotive: under-the-bonnet components

Household: hair dryer, oven, microwave oven, iron and fan heater components, coffee machines, hot water dispensers

Plumbing: snap joints

Process plant: transparent pipelines, chemical plant pumps

Textiles: very large rollers for finishing processes.

World consumption of PSUs is believed to be about 7,000 tonnes per annum. About 30% of this is of glass fibre filled grades. Manufacturers are Amoco and ICI. Products of these two manufacturers differ slightly in physical characteristics and properties.

Applications make use of PES's combination of high temperature performance and lightness. They include:

Automotive: needle and ball bearing cases in gear boxes

Mechanical: high precision/close tolerance mouldings, transparent windows and casings, hot fluid valves, heater fans, heavy duty welding units for offshore oil

Electrical: transformer bobbins and winding jackets, radomes and circuit boards

Medical: sterilizable components (e.g. for dialysis machines)

Domestic: interior spindles for thermostatic radiator valves.

Sheet material is available for thermoforming components for the aircraft sector. PESs are produced by BASF and ICI.

Until the mid-1980s the polysulphones were the biggest group per volume of production among the highly heat-stable engineering thermoplastics. They are now occupying a second place, after polyphenylenesulphides. It is anticipated that the total volume of consumption of polysulphones in market-economy countries by 1995 will reach 34.2 thousand tonnes against 10,000 tonnes in 1985, with a mean annual rate of increase in consumption in the period 1986-1995 of about 13%. Specialists predict that the rates of increase in consumption of polyethersulphones and polyarylsulphones will be more than three times as high as the mean annual rates of increase in consumption of polysulphones. The polysulphones remain one of the cheapest materials in the group of highly heat-stable engineering thermoplastics ($US 8-10/kg).

I.6.4 Polyesteretherketones

Polyesteretherketones are a group of high quality, crystalline thermoplastics developed in the 1980s. At the present time, they include polyesteretherketones (whose production was developed by ICI in the United Kingdom in 1981), polyetherketones which the same firm began to produce in 1986, and polyetherketoneketones or polyetherdiketones, introduced by the Du Pont Company in the United States at the end of 1987.

Common features of polymers in this group are the excellence of their physico-mechanical properties (strength, rigidity, impact strength) and of their heat, chemical and fire resistance. The protracted service temperatures for polymers in this group exceed 200° C. The materials are characterized by low moisture absorption and high hydrolytic stability. They are distinguished by their high degree of resistance to radiation weathering and their resistance to organic and inorganic reagents.

Polyetheretherketone (PEEK) is a crystalline material moulded at high temperatures (360-390° C). It is prepared by polyetherification methods similar to those used in the manufacture of polysulphones.

Crystallization rates are slow, allowing thinner sections to be quenched to form a near transparent material. Its very high melt viscosity made it originally a coating and wire-covering material, but moulding grades are available.

Outstanding properties are high temperature use; flame resistance; low smoke generation; chemical and solvent resistance (except for concentrated sulphuric acid); low friction; and toughness and fatigue resistance.

Mechanical properties include high stiffness at 23° C with non-brittle failure on impact. Flexural strength is high. PEEK has excellent failure resistance. Creep effects are low. However, its high viscosity, high processing temperatures, and narrow processing range make it an intractable material to mould. It is often used in glass reinforced form and as a binder to make carbon fibre based tapes and sheets.

Filled and unfilled grades of polyetherketones are produced. Among the fillers used are glass and carbon fibres and graphite. Composites are proposed with electroconductive reinforcing agents or with various additives to increase resistance to wear. The ICI Company has developed a special grade of polyetherketone known as Victrex KX3, with fibrous and powdered fillers, for making self-lubricating bearings.

The polyetherketones can be used for making film, sheet, fibre, pipes, rods or moulded products on standard equipment for injection moulding, extrusion or pressing. About 70% of unfilled polyesteretherketone and 90% of the reinforced grades are processed by injection moulding.

Important fields of application of polyetherketones are electrical engineering and electronics, where they are used for preparing insulation, including multi-layer insulating coatings, printed circuit boards, and various moulded parts. Preservation of mechanical strength under highly humid and hot conditions and resistance to various types of radiation have made application of these compounds promising for the aerospace industry. The possibility is being considered of using polyetherketones in various branches of mechanical engineering, atomic energy and the manufacture of sports goods.

Polyetherketones are now assuming great importance as thermoplastic matrices for advanced composites, particularly in the production of carbon-filled plastics, which are considered alternative materials to epoxide composites in the aerospace industry.

Main applications make use of PEEK's high temperature resistance; low friction; flame resistance; chemical resistance; and good electrical performance. It is particularly appropriate for aggressive environments. Uses include:

Automotive:	bearings, engine parts
Electrical:	cable insulation, connectors
Mechanical:	"soft" valve seats, high pressure steam valves, pump impellers, bearing linings, filter cloths
Chemical:	pumps and impellers for hot oil (as filaments), hot filtration cloths
Domestic:	hot water armatures
Aircraft/military:	components for aircraft engines and weaponry
Nuclear:	components in or near reactors.

Wire coverings are of interest because of the excellent fire properties and resistance to cut. Its properties are used in miniaturization. PEEK is produced and marketed by ICI. Polyetherketone (PEK), which is closely analogous from a chemical point of view, has similar properties and applications. It is marketed by ICI, Hoechst, and BASF.

Present world consumption of polyetherketones is estimated by specialists of the Amoco Chemical company at 770 tonnes per annum. According to a forecast by ICI, the world market for polyetherketones by the end of the 1990s will reach 1.4-2.3 thousand tonnes, and their price will fall from the present level of $US 77/kg to about $US 22/kg.

I.7 Fluorine-containing polymers

I.7.1 Polytetrafluoroethylene (PTFE)

PTFE is made commercially by two processes, one leading to a granular polymer and the other leading to a polymer dispersion of much finer particle size and lower molecular weight. Few details are available regarding the processes involved.

PTFE is a tough, flexible, non-resilient material of moderate tensile strength but with excellent resistance to heat and chemicals; it is only attacked at room temperature by molten alkali metals and in some cases fluorine. It is an excellent electrical insulator. It remains ductile in compression at temperatures close to absolute zero. The coefficient of friction is very low and is said to be lower than that of any other solid. Though it has good weathering resistance, PTFE is degraded by high energy radiation. It is not wetted by water and is "non-stick".

Normal thermoplastic processing methods cannot be used because of PTFE's high viscosity. It is usually processed by preforming and by sintering (this results in emission of toxic cyclic fluoro compounds). Granular polymer can be extruded, but this must be done very slowly.

Applications reflect PTFE's chemical inertness; good weathering resistance; good electrical insulation characteristics and heat resistance; non-adhesive properties; and very low coefficient of friction. However, uses are limited by its high cost.

Its chemical inertness leads to use in a variety of seals, gaskets, packings, valves, and pump parts in the chemical industry. Its excellent insulation properties lead to its use in wire insulation; valve holders; insulated transformers and hermetic seals for condensers; in laminates for printed circuitry; and other miscellaneous electrical applications.

Its low coefficient of friction leads to use in lining chutes or coating metal objects where low friction or non-adhesive characteristics are required. In the automotive and mechanical engineering industries, it is used for pivot bushes, piston rings, hydraulic seals, packings, gaskets, and bearings.

Because of its flexing resistance, inner linings made of dispersion polymer are used in flexible steam hose. A variety of mouldings are used in aircraft and missiles and other applications where high temperatures are encountered. It is also used for sleeves in heat exchanger equipment.

PTFE is rarely used to make large objects because of its high cost. It is often possible to coat a metal object with PTFE to meet a requirement. PTFE is made by Du Pont and ICI, and world production is about 31,000 tonnes per annum.

I.7.2 Polychlorotrifluoroethylene, polyvinyl fluoride and polyvinylidene fluoride

Polychlorotrifluoroethylene, polyvinyl fluoride and polyvinylidene fluoride are representatives of the group of fluoroplastics - special-purpose engineering materials that are highly thermostable.

The industrial production of polyvinyl fluoride was developed by Du Pont in the United States at the beginning of the 1940s. Production of polychlorotrifluoroethylene was begun in 1953 by the Japanese firm Dainin Koge and of polyvinylidenefluoride in 1960 by Pennwalt Chemical in the United States.

The fluoroplastics are usually obtained by suspension, emulsion, bulk or solution polymerization. Independently of the means of polymerization used, the products obtained have identical chemical structure and are characterized by their physical form, which determines the methods preferably used for moulding them and their fields of application. The fluorinated plastics obtained by solution, bulk or suspension polymerization are produced in the form of granules and are used mainly for moulding; those obtained by emulsion polymerization occur in the form of finely dispersed powders and are processed by extrusion and mainly used for manufacturing coatings, paints and lacquers. Some companies produce fusible granulates for extrusion, injection moulding and forming, and also water dispersions for coatings and impregnating agents.

Properties common to all the fluoroplastics are high resistance to chemical agents, maintenance of mechanical properties in a temperature range from -200° to +260° C, good dielectric and anti-adhesive characteristics, low coefficients of friction and water absorption, and a high degree of fire resistance. Polychlorotrifluoroethylene is distinguished from other fluoroplastics by the presence of chlorine atoms in the macromolecular chain. This is responsible for specific properties. Polychlorotrifluoroethylene is resistant to radiation. Its degree of crystallinity and density vary according to the type of cooling used after moulding. In most cases the degree of crystallinity is 45-65%. Polymers with a lower degree of crystallinity are quite viscous and optically transparent, while those with a higher degree of crystallinity are more rigid, possess barrier properties against gases and liquids and are less transparent.

Polychlorotrifluoroethylene film has the lowest coefficient of moisture permeability of all the transparent polymer films. The material is proof against most chemical reagents but not at a temperature above 100° C. It swells and dissolves in chlorinated solvents. The maximum temperature at which it can be used over long periods is 120° C.

Polyvinylidene fluoride is a crystalline polymer of high molecular mass and is distinguished from completely fluorinated polymers by having greater mechanical strength, wear and creep resistance and dimensional stability. It is resistant to solvents and chemicals, including such oxidants as bromine and bromate solutions. Polyvinylidene fluoride possesses piezo-electric properties. It can be used for protracted periods in a temperature range from -100° to +150° C.

Polyvinyl fluoride is a highly crystalline polymer which retains its service properties at temperatures ranging from -70° to +110° C and is characterized by good impact strength and resistance to wear and weather. It is marketed only in the form of film because of the complexity of the moulding process. Polyvinyl fluoride film has a lower gas permeability and higher strength than fluoropropylene-ethylene films.

The main fields of application of the fluoroplastics are for chemical machinery, the electronics and electrical engineering industries, the automotive industry and aeroplane construction. The structure of consumption of these materials in the market-economy countries in 1987 was as follows (%):

Field of application	United States	Western Europe	Japan
Chemical machinery	25	40	35
Electrical engineering and electronics industries	50	20	20
Automotive and mechanical engineering industries	15	25	30
Other (building technology, coatings, domestic appliances, leisure goods)	10	15	15

Sources: Kunststoffe, 1987, 77, No. 10, 1016-1019.
Kunststoffe, 1987, 77, No. 10, 49-51.

Polychlorotrifluoroethylene is marked by its good physico-mechanical characteristics at low temperatures, dimensional stability, low permeability to gases and suitability for the production of very-low-temperature valves, gaskets, and sealing compounds. Polychlorotrifluoroethylene films, which are highly transparent and almost impermeable to moisture, are used for packing precision electronic apparatus, medicaments and electro-luminescent instruments.

PCTFE monomer is produced from hexachloroethane via trichlorotrifluoroethane. Polymerization is carried out by techniques similar to those used in the manufacture of PTFE.

The chemical resistance of PCTFE is good but not as good as that of PTFE. It can be attacked by some strong acids and alkalis. It has a lower softening point than PTFE and a higher tensile strength. Its electrical insulation properties are poor. It may be processed by normal thermoplastic processing methods, but because of the high melt viscosity, high injection moulding pressures are required. It is transparent in thin films and has greater hardness and tensile strength than PTFE.

PCTFE is used less than PTFE. Applications include:

Medical and military:	gastight packaging film
Chemical:	transparent windows of apparatus and plants
Mechanical:	seals, gaskets and O-rings
Electrical:	hook-up wire and terminal insulators
Nuclear:	handling of uranium hexafluoride.

PCTFE is manufactured in Europe by Hoechst, and consumption is estimated at 350-400 tonnes per annum.

Polyvinylidene fluoride (PVdF), because of its good strength characteristics, is used in the production of circuit couplers in the computer and radio industries. The resin is used for producing contact-free switches, magnetic heads, microphones and electrodes, in the operation of which use is made of the piezo-electric and other electrical properties of the polymer. Polyvinylidene fluoride is widely used for insulating various electric signalling circuits fitted in premises. Thanks to their fire resistance and thermal stability, these fluorinated polymers do not need additional insulating tubes and this makes it possible to halve production costs. A promising trend is the production of polyamide fabrics with polyvinylidene fluoride coatings that are unidirectionally permeable to moisture, used in making sports clothes for open-air exercise.

PVdF monomer is made by dehydrochlorination of 1-chloro-1,1-dichloroethane. PVdF is one of several fluoropolymers that can be processed by injection moulding. All have excellent chemical and heat resistance, they are good electrical insulators, and they have high impact strength. PVdF has the highest dielectric constant, heat deflection temperature, flexural strength and modules. Its yield strength and creep resistance is high for a fluorinated thermoplast.

PVdF has a very wide range over which it can be processed (140° C), distinguishing it from other fluoropolymers. Abrasion resistance is similar to PA. As thermal and light stability is very good, no heat or UV stabilizers are required.

High molecular weight grades are only used for large items. The low molecular weight (crystalline) forms are generally used when products are stiff and strong (similar to PA).

PVdF is very resistant to a wide range of chemicals including halogens, salt solutions, and inorganic acids and bases. It resists degradation by UV light and alpha and beta radiation.

It is not resistant to oleum, fuming nitric acid, hot concentrated bases or alkali metals. It is swollen by polar solvents like acetone and ethylacetate. Some stress cracking occurs in hot alkalis.

PVdF has piezo-electric properties. After being heat treated and cooled in a DC field, it is three to five times more piezo-electric than quartz. This makes PVdF attractive in the manufacture of transducers and loudspeakers. PVdF also has pyro-electric properties leading to its use in pyro-electric detectors.

Because of its resistance to heat and chemicals (especially chlorine), PVdF pipes and fittings are used in the chemical and cellulose industries. Its chemical inertness also makes it useful in the semi-conductor industry; in the manufacture of ultra pure water and high purity acids; in biotechnology; and in the pharmaceutical and food industries.

PVdF film can be made in mono- or bi-axially oriented versions. It can be laminated to PP, PVC, and bitumen. This makes it useful for bitumen roofing protection. It has very good weathering properties and may be used continuously up to 150° C.

It is produced by Dynamit Nobel, Kay-Fries, Kureha, Laporte, Pennwalt, Solvay, and Atochem.

Polyvinyl fluoride (PVF) films are very strong and weather-resistant. They are used for facing solar panels, as sound-insulating material in the building industry and in the internal finishing of aircraft.

Details of the commercial method of preparing PVF monomer are not available, but it may be prepared by adding hydrogen fluoride to acetylene at about 40° C.

PVF resembles PVC in its low water absorption, its resistance to hydrolysis, and its insolubility in common solvents at room temperature. However, it has a much greater tendency to crystallize, better heat resistance, and exceptionally good weather resistance. It will also burn very slowly. Instability at processing temperatures makes handling difficult, but Du Pont has been able to manufacture and market its Tedlar film.

This is used in the manufacture of weather-resistant laminates, for agricultural glazing, and for electrical applications. It is also used as a release sheet for epoxy printed circuit boards, automotive side trim, and various aircraft cabin applications.

In absolute terms the amount of fluoroplastics consumed is small. In 1987 it was 48,000 tonnes, of which 31,000 tonnes was accounted for by polytetrafluoroethylene - the most important polymer in the group, 7,000 tonnes by polyvinyliddenefluoride and 10,000 tonnes by the other fluorinated polymers. Specialists from the Atochem company in France forecast that by 1995 consumption of fluoroplastics in market-economy countries will increase to 85,000 tonnes per annum, with the proportion of polytetrafluoroethylene reduced to 53% and that of polyvinylidenfluoride increased to 21%. The considerable extension of demand for polymers in this group is hindered by their high cost ($US 12-66/kg of foam in the United States in 1988) and the complexity of processing methods.

I.8 Acrylic plastics

I.8.1 Polymethyl methacrylate (PMMA)

PMMA monomer is obtained by reacting acetone with hydrogen cyanide to give acetone cyanohydrin. This is subsequently treated with sulphuric acid and then reacted with methanol. Sheet is prepared by polymerization in bulk.

PMMA is a crystal clear, uncoloured amorphous plastic. At room temperature it is hard and rigid, maintaining its clarity even in thick sections and after long exposure to outside atmospheres (though it is liable to surface scratching caused by airborne dust or cleaning). It has some electrical insulation properties and is used for low frequency applications, but it is inferior to polyethylene or polystyrene, particularly at high frequencies.

Polymers and copolymers based on acrylic and methacrylic monomers possess a great variety of properties, combining the best optical qualities, weather resistance and transparency to light among all polymers with excellent physico-mechanical properties, good colourability and biological inertness. The main representative of commercial acrylic resins is polymethylmethacrylate (PMMA). Data on the production of acrylic resins by the different countries are contradictory and incomparable. The estimated production of PMMA is shown below in thousands of tonnes.

Country	1985	1988	1989
United States 1/	253	317	336
Japan	147	187	196
United Kingdom	27	111	-
France	80	91	-
Italy	60	63 2/	-

Source: Japan Plastics, 1990, 41, 6, 27.

1/ Amount sold.

2/ 1986.

The largest producer of PMMA in the world is the American firm Rohm and Haas. It produces acrylic resins under the trade name of Plexiglas in three plants: in Bristol (Pennsylvania), Knoxville (Tennessee) and Louisville (Kentucky). In 1983, total productive capacity was estimated at 170,000 tonnes. According to a statement by the firm, in 1990-1991 it will extend the capacity of two of its factories by 45-50%. Other important suppliers of PMMA in the United States are the Cyro and Du Pont firms.

These three greatest producers of PMMA in the United States are the only ones to produce also methylmethacrylate - the raw material for PMMA. They use about 60% of the monomer produced themselves (50% for sheet and moulded composites and 10% for lacquers and paints); 40% goes on to the market. In all, about 50 firms in the United States produce acrylic resins and items made from them.

The main producer of PMMA in western Europe is ICI in the United Kingdom. It produces sheet by batch bulk polymerization (Perspex trademark) and by continuous extrusion (Transpex), while moulded PMMA is manufactured by suspension polymerization (Diacon).

In Germany, three firms produce PMMA: Rohm and Haas GmbH, Degussa, and Resart; their capacities are not known. In France, the Ugilor firm produces PMMA.

In Japan, the biggest producer of PMMA is the Mitsubishi Rayon firm. In volume of production this firm is second only to Rohm and Haas.

A large proportion of PMMA is obtained by bulk polymerization in the form of sheet, rods and pipes. This method, which has been used for a long time, consists in polymerizing an initiated monomer or prepolymer in a vessel bounded by two silicate glass walls with filler between them. A disadvantage of this batch process is its laboriousness and its high cost.

At the beginning of the 1970s, a continuous method of obtaining sheet PMMA was introduced in the industry. In this case, the initiated monomer is polymerized between two continuously moving parallel flat bands. The method was developed in Japan (Mitsubishi Rayon), in the United States (Du Pont and Rohm and Haas) and in other countries.

Yet another continuous method of obtaining products direct from the monomer - the extrusion method - has been developed and introduced. This method has been used since the end of the 1960s by ICI at its two plants. In Japan, at the beginning of 1990, capacity for producing extruded sheet (50.6 thousand tonnes) exceeded capacity for cast sheet (48.15 thousand tonnes).

Extruded sheet is somewhat inferior to cast sheet in mechanical strength, thermal stability and outward appearance, but is considerably cheaper. In 1979 in the countries of western Europe, 31,000 tonnes of extruded sheet was used as against 62,000 tonnes of cast sheet. The corresponding figures in 1989 were 46,000 and 90,000 tonnes.

Moulding compounds (powders, granules) are obtained by suspension polymerization.

It is widely used to replace glass since it is significantly lighter than glass, is easier to shape and work, and is considered one of the hardest thermoplastics, being about as hard as aluminium.

PMMA is resistant to dilute acids and alkalis; concentrated alkalis and concentrated hydrochloric acid; aqueous solutions of salts and oxidizing agents; fats; oils and aliphatic hydrocarbons; dilute alcohols and detergents.

It is not resistant to concentrated oxidizing acids (nitric and sulphuric) and alcoholic alkalis. It is soluble in most aromatic and chlorinated hydrocarbons, esters and ketones. It is plasticized by ester-type materials such as tritolyl phosphate and dibutyl phthalate, and is swollen by alcohols, phenols, ether and carbon tetrachloride.

PMMA components are rigid, dimensionally stable, odourless, resistant to many common chemicals and easy to decorate. They have low moisture absorption and great UV resistance. There are some grades which are softer, tougher and have better environmental stress cracking resistance (though they are more hazy than basic grades). These properties are useful if products come into contact with aqueous detergents or soap solutions. The optical properties of PMMA have been used in the development of optical fibres. Uses include:

Automotive:	rear lamp housings, coach roof lights, motor cycle windscreens, cabins for tractors and earthmovers, fascia panelling heat resistant reflectors, (in development: headlamp lenses)
Mechanical:	machine guards
Electrical:	light fittings and covers
Telecommunications:	fibre optic cables
Aircraft:	glazing for aircraft and helicopters
Medical:	covers for baby incubators, dentures, intraocular lenses
Household:	baths and washbasins, covers of solar collectors
Sports:	"Perspex" squash courts.

Consumption of acrylic resins by type of production in the United States, western Europe and Japan is shown below in thousand tonnes.

Material	1985	1988	1989
United States			
Cast sheet (including imports)	86	118	125
Moulded and extruded composites	66	91	99
Special grades (with high impact strength and other modifications)	30	36	39
Coatings	45	46	47
Other	25	25	25
TOTAL	252	316	335
Western Europe			
Cast sheet	67	87	90
Extruded sheet	36	44	46
Moulded and extruded composites (excluding sheet)	57	62	67
TOTAL (without emulsions and coatings)	160	193	203
Japan			
Sheet materials	59	67	71
Moulded composites	50	50	52
TOTAL	109	117	123

Source: Modern Plastics International, 1986-1990.

The structure of consumption of PMMA by fields of application in the United States are given below (thousands of tonnes):

Field of application	1985		1988		1989	
	Sheet	Moulded	Sheet	Moulded	Sheet	Moulded
Building	44	36	60	50	64	52
Industrial products	22	15	32	19	34	20
Consumer products	6	1	10	2	10	2
Signs and advertising	8	2	8	4	9	5
Transport	6	12	8	16	8	20
TOTAL	86	66	118	91	125	99

In recent years there has been a noticeable increase in the use of transparent plastics in building. Among all the transparent polymer materials used in the United States in 1989, according to estimates, about 60% was accounted for by acrylic polymers, as against 47% in 1979. Below are given data on consumption of PMMA by the building industry in the United States in thousands of tonnes:

Field of application	1985	1988	1989
Glazing	38	60	62
Lighting fitments	15	20	21
Sanitary ware	15	18	19
Panels and cladding	8	12	13
TOTAL	76	110	115

Acrylic plastics are used for glassing in spaces, etc. in dwellings and public and industrial buildings and for building installations with transparent walls and roofs - sports halls, swimming pools, greenhouses, car parks and shopping malls.

In its illumination-engineering characteristics, PMMA is comparable with glass and in some respects, for example transparency for ultraviolet light, is better than glass. Below are given the comparative light-transmitting properties of glass and polymethylmethacrylate materials used in the building industry.

Material	Thickness in mm	Mass per unit area, kg/m^2	Heat transfer coefficient	Light transmission %
Polymethylmethacrylate				
ordinary sheet	5	6.0	5.2	92
two-layer hollow panel	16	5.0	2.8	86
	40	7.0	2.6	85
three-layer hollow panel	16	5.2	2.4	83
	32	6.5	1.9	81
Window glass	5	12.5	5.9	90

Source: Modern Plastics International, 1986-1990.

The use of transparent plastics instead of glass in building makes it possible to reduce the weight of building fitments and the amount of labour spent in transport and assembly and to improve the heat insulation of buildings. Transparent plastics are beginning to play an important role in installations using solar energy for heating buildings, where they are used as a light and reliable light-transmitting coating for solar panels installed on roofs.

Organic glass sheet, mainly acrylic but to a lesser extent polycarbonate or polystyrene, is a basic material for glazing in aircraft (pilot's cockpits and passenger cabins in aeroplanes and helicopters). On the basis of acrylic resins, bullet-proof, electrically heated, fire-resistant and abrasion-proof types of glass are manufactured. In the United States, at the beginning of the 1980s, it is estimated that 90-100 tonnes of PMMA were used for glazing in aircraft and ships.

Suspension polymethylmethacrylate glass, obtained in powder form and processed by pressing or injection moulding, is widely used for making rear lights, side-lights and reflectors for cars. Laminated organic glass is a material consisting of two or more sheets of organic glass, glued together with elastic adhesive film. The most widely used is organic Triplex based on PMMA and polyvinylbutyral adhesive films.

An important way of improving the operating characteristics of organic glass is its physical modification, carried out by orientation, and also the rational combination of different types of organic glass in multi-layer sheet.

The range of acrylic resins available in commerce is very varied and is constantly being extended. The most widespread means of modifying acrylic resins are mixing with other polymers, copolymerization, surface modification, and filling. The range of acrylic resins available is constantly being supplemented with special acrylic materials - water-absorbent and water-soluble types, imitation marble, and types for medical purposes. Suppliers include Altulor, ICI, Enimont, Resart, Rohm and Haas, Rohm GmbH, Fudow, and Du Pont.

According to estimates, 130-140 varieties of modified materials based on PMMA were available on the world market in the mid-1980s. The Rohm and Haas firm in the United States produces 23 sorts of modified PMMA, including 9 high-impact types and 3 types with increased heat stability, 3 types with decorative qualities and 2 types with increased resistance to abrasion. The Du Pont firm in the United States produces six types of modified PMMA, two of them of high impact strength and four consisting of decorative polymers. The ICI firm in the United Kingdom produces 11 types, of which 4 are high-impact PMMA, 2 heat stable, 2 chemically stable and 2 decorative. The firm Rohm and Haas GmbH in Germany produces 23 types, 7 of them with high impact strength. Resart, also in Germany, produces nine types of modified PMMA, including five high-impact types.

The impact strength of PMMA designed for glazing in public buildings and means of transport is achieved mainly by introducing elastomers.

In such fields of application as the glazing of aircraft cabins and optical lenses, the abrasive resistance of the materials takes on particular importance. It is achieved by applying coatings (based on silicone, melamine and fluorocarbon resins) onto the surface of sheet or moulded products. Abrasion-resistant materials have been developed and put on the market by the German firms Resart, Rohm and Haas and Degussa. They all use a thin layer of polysiloxane coating.

The high-impact and abrasive-resistant material Lucite AR, made by Du Pont, is treated with polyfluorosilane. In abrasion resistance it is as good as glass; i.e. in that respect this type of PMMA is 75 times better than standard acrylic sheet.

A large number of non-flammable and self-quenching acrylic materials have been developed, in which fire-resistance is achieved by introducing halide-containing or phosphorus-containing additives.

In such fields of application as modern optical devices, the active elements in lasers, lenses for spectacles, and optical-fibre cores, can very pure polymer be used. The producer firms are working actively to develop special pure PMMA with a great uniformity of properties. The Japanese firm Mitsubishi Rayon is producing a polymer of this sort by continuous bulk polymerization. It is intended for the production of optic fibres.

Developments in optically transparent plastics have been concentrated on materials for preparing optical disc supports. For new generations of optical discs, polymer materials are being developed which combine the strong features of PMMA and polycarbonates. Copolymers of methylmethacrylate and styrene and mixtures of PMMA and polyvinylidenefluoride are considered promising.

The market growth for PMMA has been much lower than for many other major thermoplastics over the past 35 years, though industry sources suggest it is in for a revival. Production in the United States in 1987 was estimated at about 200,000 tonnes and west European production was about the same. About 55% is in the form of cast products (mainly sheet), the rest being in the form of moulding extrusion compound, with rather more being used for moulding than extrusion.

I.9 Polyurethanes (PU)

Polyurethanes are based on the reaction of isocyanate with alcohol. Four main types of PU engineering plastics are used:

- injection moulding compounds
- flexible foams
- rigid foams
- RIM injection moulding compounds.

Polyurethanes are attracting increasing interest in many branches of the economy in view of the wide possibilities of obtaining various kinds of technically valuable materials based on them. In many countries about 80% of the total volume of polyurethane consumption is accounted for by foamed plastics, but that proportion is gradually falling as a result of the more rapid increase in the use of other types of products.

Injection moulding compounds

The properties of PU moulding compositions are very similar to PA6/6, but PU absorbs only about one-sixth of the water absorbed by PA6/6 under similar conditions, giving better dimensional stability and good retention of electrical insulation properties in high humidity conditions. Resistance to sulphuric acid is somewhat better than for PA6/6. Injection moulding PUs are infrequently used, as they are twice the price of PA6 and PA6/6. Where low water absorption is required, PA11, acetals, and in certain cases PCs, are cheaper and at least as satisfactory. Injection moulding PUs are marketed by Bayer.

Flexible foams

Several processes are used to produce flexible PU foams. The most common is a polyether one-shot process using polyol, isocyanate, a catalyst, a surfactant, and a blowing agent. Commercial formulations usually include other additives, and flame retardants are becoming increasingly important in view of their inflammability and tendency to emit smoke and toxic products when burning.

Flexible PU foams are mostly used for cushioning and other upholstery materials (including automobile applications). Compared with foams from natural rubber and SBR latex, they are less inflammable and have better resistance to oxidation and ageing. PU flexible foams are also useful as "foam back", and are used to stiffen or shape softer fabric. Applications are used for car door and roof trim, quilting, carpets, shoulder pads, and coat interlinings. Other uses are paint rollers, sponges, draught excluders, in soles for orthotic footwear, and packaging for delicate equipment. Suppliers include Dow, and ICI.

Rigid foams

Rigid foams are made in a similar way to flexible foams, but use polyols and more hydroxyl groups per molecule, as well as halocarbons to reduce the viscosity of the reaction mixture.

Some rigid foams are used in sandwich constructions for aircraft and building structures, but the major interest is in the field of thermal insulation, where they compete with PS and U-F foams. However, in comparison, they are cheaper, lighter, and generally less friable. They can also be formed *in situ* and are self-adhesive to most cavity surroundings, and skins. Pipe lagging is a major application. Other uses include insulation for domestic, industrial refrigerators, display freezers, insulated food containers, flasks, and hot water boilers.

The ability of rigid PU foams to be formed *in situ* has led to its use for spinal supports and fracture casts (where it has the advantage of being removable and being much lighter than plaster of Paris). Rigid foam PUs are made by ICI, Bayer, BASF and Dow.

RIM compounds

Many applications require that a foam surface be non-porous and have a good finish. This is done by reaction injection moulding (RIM), where the reaction components are mixed turbulently in a reaction chamber next to the mould cavity and then forced into the cavity.

The process was originally developed for the car industry to produce bumpers, front and rear ends, fascia panels, instrument housings, body panels, door sills, and window encapsulations. Other uses include:

Electronic: business machine and computer housings
Household: furniture
Leisure: various sports goods

A modification of the RIM process is the RRIM process in which reinforcing polymers such as glass fibres are incorporated into the polymer. Total production of PU in 1987 was 3.7 million tonnes. In the late 1980s, flexible foam accounted for about 45% of the market in western Europe, and 49% in the United States, and rigid foams for 39% in western Europe and 29% in the United States. In the United States, it was estimated that RIM accounted for about 6% of PU products.

The development of amine-containing polyurethanes is closely connected with improvements in RIM technology, for which quick-hardening reactive systems are required, capable of ensuring high productivity. The RIM technology was developed in the mid-1960s by Bayer AG. By the mid-1970s, as the result of a joint programme by the Bayer and Krauss-Maffei firms in Germany and the Mobay Chemical Corporation and General Motors in the United States, it became widely known as a highly productive industrial process for obtaining polyurethane products. At the present time, RIM polyurethanes make up 2-6% of total polyurethane consumption both in the United States and in west European countries. The main consumer of RIM polyurethanes is the automotive industry, which is responsible in the United States, for example, for about 80% of total consumption. Below is shown the consumption of RIM polyurethanes in the United States in thousand tonnes:

Field of application	1983	1989	Mean annual increase (1984-1989) %
Automotive industry	20.40	62.50	20.5
Refrigerator manufacture	0.18	6.80	83.0
Manufacture of electronic instrument housings	0.65	4.60	38.8
Production of sports and leisure goods	0.45	2.70	34.8
Agriculture	0.32	1.10	23.6
Furniture manufacture	0.45	0.90	12.2
Building	0.14	0.70	30.8
Other	3.34	4.75	28.1
Total	25.90	84.00	27.7

Sources: Plastverarbeiter, 1985, 36, N.10, 10. Plastics Engineering, 1985, 41, N.5, 12. Chemical Industry, 1986, 109, N.2, 84.

The development of glass-fibre filling of RIM polyurethanes (the reinforced RIM process) has led to materials being obtained with a two-fold or three-fold improvement in elasticity modulus, good dimensional stability and thermostability, and a low coefficient of linear expansion. Particular interest is being attracted by lightly-foamed elastomers from which such large motor vehicle parts as wings, doors and bumper coatings, and lately integral front body units, comprising bumper coating, radiator grille, front panels and part of the wings, have begun to be produced by the RIM technique.

It is estimated that from 1969 to 1985, the mean annual rate of increase in consumption of RIM polyurethanes in the United States was about 39.4% and in the period 1984-1989 it was about 21.7%. The most rapid rate of increase is in the use of RIM integrated rigid plastic foams in refrigerators. It is estimated that this increased in 1984-1989 by about 83% per year.

A considerable increase is occurring in the proportion of reinforced RIM polyurethanes (RRIM). In the United States in 1980, that proportion was only 2% of all polyurethanes but by 1984 it had reached 40%. In the west European countries about 10% of polyurethane consumption is in the form of RRIM polyurethanes.

Total consumption of polyurethanes in market-economy countries grew between 1970 and 1985 about three-fold, to reach roughly 3.5 million tonnes. In the United States, their proportion in total plastics consumption in 1989 amounted to about 5.6%. In west European countries, consumption of polyurethane in 1988 and 1989 amounted to 1,422 and 1,508,000 tonnes, respectively. Japan produced 289,000 tonnes and 308,000 tonnes of polyurethane foams in 1988 and 1989, respectively. The structure of consumption of polyurethanes in the United States is shown below in thousand tonnes:

Product and field of application	1985	1988	1989
Polyurethane foams			
Flexible, total	603	771	786
Furniture production	238	298	318
Transport	150	182	191
Manufacture of mattresses	70	81	65
Manufacture of carpet underlays	59	146	146
Lining of fabrics	-	-	*
Other	78	64	66
Rigid, total	337	428	393
Building	191	236	205
Production of refrigerators	54	77	73
Transport	30	21	22
Insulation of industrial equipment	27	43	41
Packaging	20	30	31
Other	15	21	21
Elastomers	112	132	147
Glues and other adhesives	-	)	)
Coatings	-	)138	)150
Total	-	1 469	1 476

Source: Modern Plastics International, 1987, 1989, 1990, 17, 19, 20, N.1.

* Included in "other".

According to forecasts, the demand for polyurethanes in the next 10 to 20 years will increase. The utilization of rigid foamed plastics as heat-insulating materials in building is promising. By the beginning of the 1990s, polyurethanes will occupy a considerable place in motor vehicle construction as materials for seats, body panels and bumpers. In the United States, for example, already in 1985 over 55% of all motor vehicles were equipped with polyurethane-coated bumpers. In Japan, 2 million such bumpers are produced every year. Also in favour of polyurethanes is the fact that they are an energy-saving material, both at the production stage (the reaction by which they are produced is exothermic) and at the utilization stage (polyurethane products have a low thermal conductivity coefficient and can be used for insulation). The continuous improvement of the chemistry and technology of polyurethane production and of the quality of materials based on polyurethanes is constantly bringing about changes in the structure of the consumption of these products, opening up possibilities for completely new fields of application.

Progress in the chemistry and technology of polyurethanes in the last few years has made it possible to expand considerably the choice of starting products. While earlier practically the only isocyanates used were toluylenediisocyanate (TDI) - in various ratios of the 2,4 and 2,6 isomers - and diphenylmethanediisocyanate (MDI), producers now have available various derivatives of those products obtained by chemical modification and containing urethane, allophanate, urea, isocyanurate and carbodiimide groupings. Use is also being made of hexamethylenediisocyanate, naphthalenediisocyanate, and xylylenediisocyanate. The range of polyols and cross-linking agents used has increased. The selection of polyols includes compounds with different hydroxyl numbers, molecular mass and degree of branching. Simple polyols are obtained by alkoxylation of propylene or ethylene oxides of hexanetriol, glycerine, pentaerythritol, saccharose, sorbite, ethylenediamine, triethanolamine, toluylenediamine and diphenylmethanediamine. Complex polyols are produced from adipic, phthalic, butyric and other acids and ethylene glycol, diethylene glycol, 1,4-butanediol, 1,6-hexanediol and trimethyolpropane.

Manufacturers are beginning to use cheap polyether polyols obtained by processing by-products of polyethyleneterephthalate. So-called polymer polyols have also appeared - graft polyether polyols used to confer increased bearing strength on flexible polyurethane foams. A considerable increase in thermal stability, together with a simultaneous decrease in moulding time has been achieved by substituting aromatic amines, isocyanurates and urea derivatives for glycols as cross-linking agents.

The range of catalysts used is being extended and includes some with specific effects. For example, regulation of the reaction for the formation of fire-resistant polyisocyanurate plastic foams has become possible through the introduction of trimerization catalysts. The use of highly active catalysts has promoted the development of cold-hardening polyurethane foams and highly reactive systems for processing by reaction injection moulding (RIM).

Among the most topical and promising trends in work on polyurethanes have been the transition in production to preferential use of MDI instead of TDI, the development of autocatalytic systems, the development of stronger and more thermostable polymers containing urea bonds and an increase in the fire-resistance of polyurethane foams.

The gradual switch in the production of polyurethanes from the use of TDI as the basic type of isocyanate to MDI, or more precisely, polymeric MDI (PMDI), was due, in the first instance, to a desire to reduce atmospheric pollution in the workplace, since PMDI has 15 times lower volatility than TDI. In addition, the technology for producing polyurethane foams on the basis of TDI copolymers includes an obligatory stage of prepolymerization in which toxic TDI fumes are evolved. For that reason, even though TDI is more costly, as was the case, for example, until recently in the United States, its application is rational because of the savings made in ventilation equipment.

The utilization of PMDI, with the higher functionality and greater reaction capacity that it confers, has had a favourable influence on the development of the technology for producing polyurethane foams. The use of PMDI also makes it possible to obtain materials with a wider range of tailor-made properties, because of the greater possibilities of varying the molecular structure of the polymer, in particular by chemically modifying PMDI and changing the functional mixture of isocyanates.

In the production of flexible polyurethane foams, the development of MDI-based technology has been a significant achievement, which has led to the appearance of qualitatively new materials and to greater changes in the structure of production. In the hot-hardening technology based on TDI, flexible polyurethane foams with an apparent density of less than 50 kg/m^3 were traditionally produced mainly in the form of blocks, from which products of the required shape were then cut out. The production of moulded items was economically unjustified because of the great length of the moulding cycle (10-30 minutes) and the need for subsequent heat treatment for final hardening.

As a result of the development of cold-hardening systems using MDI, the process of moulding products has been considerably simplified. The mould temperature has been reduced from about 95° C to about 50° C and the length of the moulding cycle has gone down to 3-6 minutes. The use of the new catalysts has made it possible to do away completely with after hardening of the product. Capital investments in equipment have been reduced by 50% and the energy requirements of production have been brought down considerably. At the same time, an improvement has been achieved in the properties of the foamed plastic and its flexibility has been particularly enhanced at lower density.

This technology was used later as a basis for producing all-plastic (foam plastic) motor-vehicle seats and soft furnishings. The automative industry is a large user of flexible polyurethane foams and the development of the technology of cold-hardened grades was connected to a considerable degree with the needs of the automotive industry for highly elastic materials with good bearing strength.

A new stage for the improvement of the production of flexible polyurethane foams was the above-mentioned process of using PMDI isocyanates alone. This has resulted in a further increase in productivity and technological excellence and an improvement in the quality of the products, including the ratio of flexibility to density.

A switch from TDI to MDI/PMDI and later to PMDI alone is occurring also in the production of rigid polyurethane foams, mainly because of the lower volatility of PMDI.

In west European countries, most producers of domestic refrigerators using foam plastic insulation have switched to using rigid polyurethane foams based on PMDI. In the United States this process is only just beginning. A few years ago, many firms switched over to the development of less volatile systems based on polymeric MDI. Research was carried out simultaneously both on isocyanate components and polyol components. Originally, PMDI-based materials were of lower quality than TDI-based. Their heat conductivity coefficient was higher, the moulding cycle longer, compression strength lower, and dimensional stability on cold ageing was worse. A solution to this problem was found by modifying PMDI and synthesizing new, low-viscosity polyols capable of compensating for the low flow inherent in PMDI-based systems and of providing the same or even better physical properties as when TDI is used.

The first polyols used for producing rigid polyurethane foams were polyester polyols which were soon replaced by cheaper, less viscous and more functional polyethers with a high content of primary hydroxyl groups.

Yet another important trend in the development of polyurethane chemistry should be noted - the introduction of an increasing number of urea bonds into the polymer structure, going as far as their complete substitution for urethane bonds. This is leading to increased strength and thermostability. This trend is due to the requirements of the automotive industry, which needs strong and thermostable materials for producing body panels by the RIM method. The thermostability of polyurea resins is better suited to the conditions faced by the material during production of motor vehicles, for instance the effect of high temperatures during paint work.

In recent years, there has been a switch from mixed polyurethane-polyurea materials to completely polyurea materials. These materials are the result of the consistent development of polyurethane chemistry and, at present, the largest firms producing polyurethanes also have their own developments of polyurea products.

I.10 Styrene copolymers

I.10.1 Styrene acrylonitrile copolymer (SAN)

Copolymers of styrene and acrylonitrile, in a ratio of about 70:30 by weight, are rigid, transparent thermoplastics in many ways resembling polystyrene, but with additional desirable qualities. They have a higher softening point (about 95° C) and are harder; more rigid; more craze resistant and resistant to environmental stress cracking; and more solvent resistant. SAN has higher resistance to creep under load than PS, and a higher heat distortion temperature, but still has poor impact strength. SAN has better chemical resistance than PS and, like PS, is a good electrical insulator. Maximum continuous use temperature is about 85° C. This can be raised by about 10° C by adding glass fibre by 35%. This produces a very rigid material. SAN is transparent, with a pale yellow colouring which can be offset by the use of a blue dye.

SAN is resistant to saturated hydrocarbons; low aromatic engine fuels and oils; animal and vegetable fats; oilsaqueous salt solutions; and dilute acids and alkalis. It is not resistant to aromatic and chlorinated hydrocarbons, nor to esters, ethers, ketones, most chlorinated hydrocarbons, or concentrated inorganic acids. Uses include dials, knobs and covers for domestic appliances, electrical equipment and car equipment; picnicware and housewares.

SAN is produced by BASF, Dow, Enimont, and Monsanto. Production has been estimated at about 50,000 tonnes per annum in the United States and about the same in western Europe. SAN copolymers have tended to be superseded by ABS copolymers, which were developed to remedy some of their defects.

I.10.2 Acrylonitrile butadiene styrene copolymer (ABS)

ABS plastics are usually made by polymerizing styrene and acrylo-nitrile into polybutadiene, in a polybutadiene latex. The resultant grafted polybutadiene phase is then melt compounded with SAN, which comprises 70%+ of the total composition.

By varying the monomer ratios, the way in which they are combined, the size and proportion of rubber particles, the cross-link density of the rubber particles, and the molecular weight of the SAN, a wide range of materials can be produced differing in impact strength and ease of flow. As a general rule, as the molecular weight of the SAN is increased, the strength and rigidity of ABS increases and as the rubber content increases, the strength, hardness, heat resistance and rigidity of the ABS decreases. ABS can be divided into injection moulding and extrusion grades (the former having much lower viscosity). Each of these can be divided into medium, high and very high impact grades. There are also high heat, plating and flame retardant grades.

ABS is generally a hard tough material with good impact resistance even at low temperatures. It has low water absorption and is a good electrical insulator. Electrical properties are unaffected by changes in humidity. Mouldings are dimensionally stable and high gloss is required. The surface is resistant to scuffing but the material has poor weathering properties. ABS has better heat resistance and impact strength than toughened PS, higher flexural moduls than PP, and is not as notch sensitive as PC and PA. It is easily electroplated after acid etching.

ABS is resistant to alkalis, acids other than concentrated oxidizing acids, salts, oils and fats, most alcohols, and hydrocarbons.

The ABS copolymers occupy an important place among engineering plastics. In scale of production, these copolymers are among the highest tonnage plastics. Data on the production of ABS copolymers in the United States and Japan are shown in table 1. For other countries, statistical data on the production of these resins are not given separately and are usually included in total production of styrene resins. In 1983, the world production of ABS copolymers was 1.6 million tonnes and it is estimated that in 1990 it will be 1.75 million tonnes. The United States occupies the first place for total production of ABS copolymers. However, it is in Japan that their production is increasing most rapidly. Thus, in the period 1981-1988, the mean annual rate of growth in ABS copolymer production was 3.9% in the United States and 8.45% in Japan. Since 1980, Japan has held the leading place for production of ABS copolymers per capita.

Table 1. Production of copolymers in the United States and Japan

Country	1965	1970	1975	1980	1985	1988	1989
United States							
Production, thousand tonnes	186	258	303	417	480	563	565
Proportion of total plastics production, %	3.5	3.0	2.8	2.5	2.1	2.2	2.1
Production per capita, kg	0.9	1.3	1.4	1.8	2.0	2.3	2.3
Japan							
Production, thousand tonnes	12	138	150	260	421	498	508
Proportion of total plastics production, %	0.8	2.7	3.0	3.4	4.6	4.5	4.2
Production per capita, kg	0.1	1.3	1.4	2.2	3.8	4.0	4.1

Sources: Modern Plastics International, 1970-1990, 1-20, N.1. Plastics modern elastomer, 1970-1990, N.1-2. Chemical Market Reporter, 1990, 237, N.7,10.

General information on the capacities of plants for producing ABS copolymers is shown in table 2. As will be seen, in the period 1980-1985 in a number of countries, such as the United States, United Kingdom, Spain, Italy and France, the closure of obsolete plants and the reconstruction of others caused a reduction in the production capacities of ABS copolymers. However, in some countries in 1988, as compared with 1985, capacities of plant for making these copolymers had increased. The changes in the coefficient of utilization of capacity in plants for producing ABS copolymers in the United States and Japan (%) are given below:

Country	1970	1975	1980	1985	1988	1989
United States	67.9	52.0	52.0	71.0	79.0	76.0
Japan	75.8	47.3	78.0	93.5	82.0	-

Table 2. Plant capacities for the production of ABS copolymers in certain market-economy countries

Country, indicator	1970	1975	1980	1985	1988
United States					
Number of plants	11	12	12	11	10
Capacity, thousand tonnes/annum					
total	380	583.2	800	680	710
mean	34.4	48.6	77	66.2	71
maximum	90	131.4	-	-	-
Belgium*					
Number of plants	-	1	1	1	1
Capacity, thousand tonnes/annum					
total	-	50	75	90	90
mean	-	50	75	90	90
maximum	-	50	75	90	90
United Kingdom					
Number of plants	4*	2	4	1	1
Capacity, thousand tonnes/annum					
total	42*	35	120	70	60
mean	10.5*	17.5	30	70	60
maximum	20*	20	70	70	60
Spain					
Number of plants	-	1	4	4	3
Capacity, thousand tonnes/annum					
total	-	10	53	44	49
mean	-	10	13	11	16
maximum	-	10	20	20	30
Italy					
Number of plants	4	4	3	2	2
Capacity, thousand tonnes/annum					
total	35	52	89	70	65
mean	8.8	13	27	35	32
maximum	15	24	40	-	35
Netherlands*					
Number of plants	1	3	3	3	5
Capacity, thousand tonnes/annum					
total	25	88	152	160	190
mean	25	29	48	53	38
maximum	25	50	70	70	70
Germany*					
Number of plants	3**	3	2	2	2
Capacity, thousand tonnes/annum					
total	80**	150	165	175	225
mean	27**	50	83	87	112
maximum	-	75	90	100	150

Table 2. (continued)

Country, indicator	1970	1975	1980	1985	1988
France					
Number of plants	4	4	2	1	1
Capacity, thousand tonnes/annum					
total	38	68	55	50	50
mean	9.5	16	27	50	50
maximum	25	40	35	50	50
Japan					
Number of plants	12	12	12	13	10
Capacity, thousand tonnes/annum					
total	182.4	317.4	330	450	608
mean	15.2	25	27.5	35	60
maximum	31.2	48	60	70	72

Source: Information Chimie, 1988, N.294, p.182

* Including SAN-copolymers (styreneacrylonitrile copolymers).
** 1971.

Suppliers include Atochem, BASF, Bayer, Dow, DSM, Enimont, Monsanto, Rhone Poulenc, USS Chemicals, and G.E. Plastics. West European production was estimated to be 485,000 tonnes in 1989. An increase of nearly 20% has been forecasted by 1993. This means that production is only exceeded by that of bulk polymers.

Beginning in 1980 there has been a tendency to increase the degree of capacity utilization in plants for making ABS copolymers, and at the present time the ratio is at a fairly high level. The largest firm for production of ABS copolymers is General Electric. Its share in total capacity for ABS copolymer production in the United States was 46% in 1990. The share of the next biggest company, Monsanto, was 34%.

In 1988, these copolymers were produced at one Monsanto plant in Antwerp (Belgium) with a capacity of 30,000 tonnes, at a Borg Warner Chemicals plant in Grangemouth (United Kingdom) with a capacity of 60,000 tonnes, in two Enichem Technoresins company plants in Italy (at Ferrara and Ravenna) with a combined capacity of 65,000 tonnes, in the Netherlands in three plants, one in Amsterdam (Borg Warner Chemicals Europe with a capacity of 70,000 tonnes), one in Ternezen belonging to the Dow Chemical Company with a capacity of 50,000 tonnes and one in Geela belonging to the DSM Company with a capacity of 50,000 tonnes. In Germany, they were produced in two plants of the BASF and Bayer firms with a combined capacity of 275,000 tonnes and in France, at a single plant belonging to Borg Warner Chemicals with a capacity of 50,000 tonnes. In Japan, ABS copolymers are produced by ten firms in ten plants with a total capacity of 608,000 tonnes. The share of the largest firms - Asahi Kagaku and Nippon Gosei Rabe - in 1988 was 29.8%.

On an industrial scale, ABS copolymers are obtained by emulsion, suspension or bulk polymerization. In addition, a mechanical process is used for mixing SAN-copolymers with butadiene-acrylonitrile rubber. The method chosen depends on the intended use of the product and on its properties.

Most of the ABS copolymers being made at present are a mixture of a graft copolymer of acrylonitrile and styrene in polybutadiene with SAN-copolymers. The industrial grades usually contain 50-90% of SAN-copolymer and the ratio of acrylonitrile to styrene is, as a rule, 25-30:70-75. Graft copolymers possess higher impact strength than mechanical mixtures of SAN-copolymers with rubber.

Consumption of ABS copolymers in market-economy countries in 1985 was 1,600,000 tonnes and in 1990 and 1996 (estimated) it should reach 1,750,000 and 2,370,000 tonnes respectively. Data on ABS copolymer consumption in some countries are given in table 3. Japan occupies first place in per capita consumption of these copolymers.

Table 3. Consumption of ABS copolymers in selected countries

Country	1985	1988	1989	1990 (Estimate)
United States				
Total, 1,000 tonnes	434	459	501	-
Per capita, kg	1.7	1.8	1.9	2.0
United Kingdom				
Total, 1,000 tonnes	57	-	-	62
Per capita, kg	1.0	-	-	1.1
Italy				
Total, 1,000 tonnes	60	-	-	67
Per capita, kg	1.1	-	-	1.2
Germany				
Total, 1,000 tonnes	118	-	-	125
Per capita, kg	2.0	-	-	2.1
France				
Total, 1,000 tonnes	66	78	80	70
Per capita, kg	1.2	1.4	1.4	1.3
Japan				
Total, 1,000 tonnes	380	416	462	534
Per capita, kg	3.1	3.4	3.7	4.1

Data on the structure of ABS copolymer consumption in certain countries are given in Tables 4 and 5.

Table 4. Structure of ABS copolymer consumption in the United States

Field of Application	1988	1989
Total, thousand tonnes	459	501
Including, %		
Building	15.3	14.5
Parts for:		
- the automotive industry	20.7	20.0
- appliances	20.0	19.5
- calculating machines and telephones	8.9	9.5
Packaging	0.4	0.3
Electrical engineering and electronics	9.2	9.0
Bags and cases	1.3	1.4
Modifiers	2.8	2.8
Furniture parts	0.4	0.5
Other	21.0	22.5
TOTAL	100.0	100.0

Source: Modern Plastics International, 1970-1990, 1-20, N.1.
Plastique Moderne et Elastomère, 1970-1990, N.6.
Promt, 1988, 80, N.7, p. 51.
Promt, 1988, 80, N.2, p. 49.
Polymer News, 1989, 14, N.1, p. 22.
Plastics Industrial News, 1988, 34, N.10, p. 149.

The automotive industry is one of the largest and most promising fields of application of ABS copolymers in all the countries with a market economy. In 1989, in the United States, consumption for these purposes was 100,000 tonnes, in the countries of western Europe it was 122,000 tonnes and in Japan it reached 102,000 tonnes. In Japan, an important place in the consumption of these resins is occupied by the production of domestic electrical appliances (27.7% of total consumption of ABS copolymers in 1989). Wide use is also made of these resins by such leading car manufacturers as Renault, Fiat, Volkswagen, Ford, Citroen and General Motors. The resins are used for making many external and internal automobile parts, radio housings, instruments, cooling and heating systems, tool boxes, arm rests, radiator grilles and gearbox casings. The use of ABS copolymers to manufacture domestic electrical appliances (housings and other parts of tape recorders, radio sets, television sets, refrigerators, dishwashers and washing machines) is a relatively important sphere of utilization of these resins and, for example, in Japan and the United States in 1988, represented 118,000 and 92,000 tonnes, respectively.

Table 5. Structure of ABS copolymer consumption in western Europe (other than France), France and Japan

Field of application	1985			1988			1989		
	Western Europe	France	Japan	Western Europe	France	Japan	Western Europe	France	Japan
Total, 1,000 tonnes	406	66.0	380	466	78.4	416	485	80	462
Including, %									
Parts for domestic electrical appliances	19.7	22.7	34.5	16.7	33.7	28.4	17.7	34.4	27.7
Parts for the automotive industry	24.6	37.9	20.8	25.8	40.8	22.8	25.2	37.5	22.0
Household goods	22.8	17.1	19.7	20.2	11.7	21.9	17.3	10.5	22.0
Industrial machinery	*	*	21.6	*	*	24.7	*	*	25.5
Other	32.9	22.3	3.4	37.3	13.8	2.2	39.8	17.6	2.8
Total	100.0	100.0	100.0	100.0	100.0	100.0	100.0	100.0	100.0

Source: The same as for Tables 3 and 4.

* Included under "other".

A rapidly expanding branch for the utilization of ABS copolymers is the production of computer appliances (calculators, meters, copying machines). Thus in 1988, in western Europe, 21,000 tonnes of ABS copolymers were used for that purpose. Alloys of ABS copolymers with polycarbonates are promising in this respect.

The consumption of ABS copolymers in electrical engineering and electronics in the United States was 53,000 tonnes in 1988 (including electrical equipment for motor vehicles). In this sphere, ABS copolymers are used for manufacturing parts and units for information system instruments, telephone apparatus and hearing aids. Glass-fibre-reinforced grades of the copolymers, with increased rigidity and heat stability, are used in informatics and telecommunications. A considerable amount of ABS copolymers are used in building in the form of pipes, profiles and sheet. However, in this sphere there is competition from polyvinylchloride, the lower cost of which has led to its squeezing out the ABS copolymers. The proportion of ABS copolymers in building fell in the United States in 1970-1989 from 24.9 to 14.5% and by 1992 it is estimated that there will be a considerable further reduction in their use and increased substitution of cheaper polyvinylchloride for them. Domestic prices for ABS copolymers in the United States in $US per tonne are given below.

Type of copolymer	1988	1990
ABS copolymer		
medium impact strength	1 807-1 917	1 983-2 093
high impact strength	1 873-1 983	2 093-2 314
very high impact strength	1 895-2 049	2 204-2 314
thermo-stable high-impact-strength copolymers	2 248-2 424	2 578-2 644
ABS copolymers for tubes		
general-purpose	1 653-1 719	1 983-2 204
high impact strength	1 719-1 785	-
for fittings	1 719-1 807	2 093-2 204
metallized	2 049-2 248	2 204-2 314
fire-resistant	2 314-2 644	2 651-2 865
structured foam plastics	1 983-2 226	2 314-2 358
glass-fibre-reinforced with a glass fibre content of		
10%	1 917-2 226	3 085-3 526
30%	1 961-2 270	2 997-3 438
Alloys of ABS copolymers		
with polycarbonates	2 226-3 063	3 210-3 306
with polyvinylchloride	1 985-2 446	3 085-3 210
with polyamides	3 746-3 857	3 901-4 385

Source: Plastic Technology, 1990, 36, N.6, p. 141.

CHAPTER II. MODIFICATIONS OF ENGINEERING PLASTICS

The first part of this chapter considers the modification of engineering plastics by using additives. The second part deals with the production of polymer/polymer blends and alloys. Composite materials are rarely used on engineering plastics, and are here ignored.

II.1 Use of additives

The properties of a polymer can often be altered considerably by incorporating additives. In some cases, additives are used for a wide range of polymers, in others they are polymer specific. The following paragraphs list the main purposes for which they are used, together with some of the main non-specific additives under each heading.

Fillers

There are two main types of filler used for engineering plastics: particulate and fibrous. Particulate fillers include calcium carbonate, china clay, talc and barium sulphate. For each of these types of fillers, various grades may be available differing in particle size and distribution; in particle shape and porosity; in chemical treatment of the surface of the filler; and in degree of purity. Ballotini, or microscopic glass balls, have a beneficial effect on the electrical properties of some engineering polymers, particularly polyamide and PMMA.

Fibrous fillers include wood flour, cotton flock, macerated fabric, macerated paper, and short lengths of synthetic organic fibres (e.g. nylon), but these are of little importance for engineering plastics. More important are glass fibre, chopped carbon fibre, and Kevlar and whiskers (single crystals of high length/diameter ratio with very high strength), which are used for specialized purposes. Asbestos should also be mentioned here, although, in view of its toxicity, its use has been virtually abandoned in the West.

In order to function, a filler must adhere to the plastic in which it is used. If adhesion is weak a coupling agent is used. For example, calcium carbonate may be coated with stearic acid. Other examples are methacrylato-chrome chloride and silane coupling agents for use with glass fibres. Recently, an interest has been shown in titanium coupling agents.

Glass fibre is by far the most common additive. It is estimated that among some of the more common engineering plastics, most grades of PPS are glass-fibre filled, about 90% by weight of the PET used is glass-fibre filled, and 30-40% of polyamide.

There are two main reasons for the use of glass fibre in this way:

- it often adds strength and toughness to the polymer, and
- it is much cheaper than any engineering polymer, so use of glass fibre can result in big savings in cost.

Plasticizers and softeners

Plasticizers are important for vinyl based and acrylonitrile polymers, but are occasionally used elsewhere. Most plasticizers are polymer specific.

Lubricants

These are of three main types:

(i) Materials which reduce the friction of mouldings when they are rubbed against adjacent materials. Examples are graphite and molybdenum disulphide, added to polyamides and other thermoplastics used in gears and bearings. PTFE is an important non-specific friction reducing lubricant. Carbon fibres also have lubricating properties.

(ii) External lubricants, i.e. materials which during processing exude from the polymer composition to the interface between the molten polymer and the metal surface of the processing equipment. Most are used with PVC and related polymers, but silicone lubricants are an important non-specific external lubricant.

(iii) Internal lubricants promote the flow of the polymer in the melt but have little effect on solid state properties. Most are used with PVC and related polymers.

Anti-aging additives

Most properties of plastics change over time, usually adversely. The main kinds of deterioration affecting engineering plastics are: chain scission (leading to loss of strength and toughness), cross-linking (causing hardening, embrittlement and changes in solubility), and discoloration. These changes result from chemical reactions of which the most common are oxidation, dehydrochlorination, and UV attack, so the main anti-aging additives are anti-oxidants, dehydrochlorination stabilizers and UV absorbers. Each will be considered in turn.

Antioxidants are of two main types, preventive and chain breaking. Amines and phenol antioxidants are examples of the latter (phenols being more common for plastics, though some paraphenylene-diamine derivatives are also used). Toxicity is a problem in some formulations. Preventive antioxidants include some mercaptans, sulphonic acids, and some zinc salts. Dilauryl carbonate is important for polyester engineering polymers.

Stabilizers against dehydrochlorination are used in chlorine-bearing polymers allied to PVC.

UV absorbers absorb primarily in the UV range. Absorption in the visible range should be negligible to avoid discoloration. The most important UV absorbers available commercially are o-hydroxybenzophenones, o-hydroxyphenylbenzotriazoles, and salicylates, which appear to function through the conversion of electrical energy into heat.

Flame retardants

These may function by one of four mechanisms:

(i) they may interfere chemically with flame propagation;

(ii) they may produce incombustible gases which dilute air supply;

(iii) they may act, decompose or change state endothermically, thus absorbing heat;

(iv). they may form an impervious fire-resistant coating preventing oxygen from reaching the polymer.

Phosphates are the most important type of fire retardants. These include tritolyl phosphate, trixylyl phosphate, and halophosphates such as tri(chloroethyl)-phosphate.

Halogen-containing compounds such as chlorinated paraffins, are also important. Bromine compounds, tribromotoluene and pentabromophenyl allyl ether tend to be more powerful than chlorine compounds.

Antimony compounds, particularly antimony oxide, are by themselves rather weak retardants but enhance the action of phosphorus and halogen compounds. Titanium dioxide, zinc oxide, molybdenic oxide and zinc borate have been used instead of, or in partial replacement of, antimony oxide.

Red phosphorus has proved to be an effective fire retardant for polyamides (see Chapter III for a more detailed discussion).

Colorants

These are very numerous, and it would be impossible to give a full review here. Many are not polymer specific, others are. Some colorants may adversely affect properties such as oxidation resistance and electrical insulation behaviour, and cannot be used where these properties are important.

Blowing agents

These are used to produce cellular (i.e. foam) versions of plastics, polyurethanes being the most important among engineering plastics even though nylons are also used in foam form. Blowing agents are for the most part polymer specific.

Additives to promote electrical conductivity

Electrical conductivity may be enhanced by the addition of carbon fibres, carbon black, metallized glass, nickel-coated carbon fibres, stainless steel fibres, and aluminium flakes. They may also be used to prevent electrostatic discharge in plastic parts. They are mostly polymer non-specific.

Impact modifiers

These are mainly for PVC, but can be used for related plastics and for certain other engineering plastics, especially acrylates. They are polymer non-specific.

II.2 Polymer specific additives

Polycarbonates

Comparatively little information is available on the use of additives in commercial grades of polycarbonate. Pigments, heat and UV stabilizers, blowing agents, and fire retardants are used but the range of materials available is very restricted due to the high processing temperatures involved.

Benzophenone and benzotriazole UV absorbers are mainly used, but are only effective if the PC composition is treated to become slightly acidic. Very small amounts (0.005%) of metaphosphoric acid, boron phosphate and phenyl neopentyl phosphite may also be used.

PTFE, silicone resins, and glass or carbon fibres are added to lubricants. Addition of carbon fibre to PC can increase flexural strength three times, and flexural moduls seven times.

Glass-fibre surfaces must be treated with a coupling agent to promote adhesion with PC. Beta-3,4-epoxycyclohexylethyl trimethoxysilane is commonly used for this purpose.

Nylons

Copper salts in conjunction with halides are the main additives used in heat stabilizers. Phosphoric acid esters, phenyl-beta-naphthylamine, mercaptobenzothiazole and mercaptobenzimidazole are commonly used. Light stabilizers include carbon black and various phenolic materials.

Plasticizers are rarely used, but Santicizer-8, a blend of o- and p-toluene ethyl sulphonamide, is marketed for this purpose.

Lubricants used are mostly graphite and molybdenum sulphide, but ethylene bis(stearamide) and similar compounds are used to enhance flow and mould release.

Polyamides are self-nucleating agents. Seeding with PA 66 and PET has been carried out to enhance the process for PA 6.

Halogen compounds and red phosphorus compounds are normally used as flame retardants.

Glass reinforced PA compounds are widely used as fillers (see Chapter I). Interest has been shown in using particulate mineral fillers.

Polyamides are normally impact resistant, but impact with sharp corners can lead to brittle failures. These can be avoided by the addition of dispersions of rubbers, polyurethanes, acrylates, and methacrylates or ABS.

Electrical properties may be improved by using ballotini (see above).

Polyesters

There is little information available on additives specific to polyesters.

Du Pont's Rynite nucleating agent is PET nucleated with an ionomer containing a plasticizer (thought to be n-pentyl glycoldibenzoate), and marketed in glass-fibre filled form.

PBT has been impact modified by adding polybutadiene rubber and cross-linked acrylic materials, which also improve the aging properties.

Polyacetals

Acetal polymers are probably never supplied for use without the incorporation of additives. Acid acceptors are used to absorb traces of acidic materials which attack the acetal linkage. Epoxides, nitrogen compounds and various basic salts have been used for this purpose.

UV stabilizers are also necessary as acetal resins are degraded by UV radiation. Carbon black, titanium dioxide, and benzophenone derivatives are used, the choice resting on the colour of the plastic.

In anti-friction lubricants, small amounts of molybdenum disulphide or larger amounts of PTFE (-> 25%) are used to reduce the coefficient of friction in bearing applications. Lower-cost alternatives include graphite and "chemical lubricants" of undisclosed composition.

Other polyphenylene derivatives

PPS grades of lubricantss are available containing typically about 15% PTFE and sometimes about 2% silicone.

Fluorine-containing polymers

Because of high processing temperatures, only inorganic colorants such as cadmium compounds, iron oxides, and ultramarines may be used with PTFE.

Electrical insulation properties of PTFE may be improved by adding alumina, silica, and lithis, which also improve dimensional stability. The latter property may also be enhanced by the addition of molybdenum disulphide and graphite. Magnetic properties are enhanced by the addition of barium ferrite.

The incorporation of titanium dioxide increases the dielectric constant, and certain boron compounds increase resistance to neutron bombardment.

Acrylic plastics

Apart from colorants, the following additives are important:

- Plasticizers: usually about 5% dibutyl phthalate;

- UV absorbers: those used include phenyl and methyl salicylate, 2,4-dihydroxybenzophenone, resorcinol monobenzoate, and stilbene;

- Fillers: ground sand is used as a filler in tiles and sanitary ware. Ballotini are also used to increase electrical properties.

Polyurethanes

Additives used which are of commercial importance for flexible foams include:

- Volatilizers: such as fluorotrichloromethane;

- Cross-linking agents: such as glycerol, pentaerythritol, and various amines;

- Anti-aging additives: such as tetravalent tin compounds, mercaptans, and organic phosphites;

- Cell size regulators: such as dimethylformamide, lecithin, and water-soluble silicone oils;

- Flame retardants (important because of the flammability of untreated polyurethane and the toxic smoke and fumes emitted when burning): phosphate compounds and antimony compounds are the most important. CM (combustion modified) foams are also produced by using aluminium hydrate, melamine, and graphite additives.

Surfactants are used to decrease the surface tension of the system and to facilitate the dispersion of water in the hydrophobic resin. They also help nucleation and stabilize the foam and control cell structure. Formulations for rigid and semi-rigid foams are similar.

While strictly speaking they are not additives, mention should be made to the isocyanates used in the reaction process, and to the need to use blowing agents (until very recently chlorinated fluorocarbons).

Styrene copolymers

Modification of ABS is normally achieved by blending, but fire retardant additives based on bromine compounds (particularly octadibromodiphenyl) have been important in the past - less so now in view of the concern about toxic decomposition products.

II.3 Polymer blends

No distinction is made here between polymer blends and polymer alloys or copolymers. While the distinction is important theoretically, it is irrelevant to the purposes of this study.

The production of completely new polymers in commercial volumes has in recent years been very limited, but the development of polymer blends has been highly significant. Particularly important have been blends combining a glassy or near-glassy resin with an elastomer or rubbery polymer. Such pairings often combine great rigidity with toughness and impact resistance in a way which is comparatively rare among unblended polymers. SAN and ABS, described in Chapter I, are the best known examples.

The number of polymer blends achieved in the laboratory is very large. Only those blends which are commercially important are examined here. Many blends have been developed for use in the automobile industry. Where no specific use is given - as is often the case in the literature - it can reasonably be assumed that automotive uses are the most important. Another important client for polymer blends is the aircraft industry.

Blends based on polycarbonates

Polycarbonate alloys with ABS and MBS (in a proportion of 2-9%) reduce the PC's notch sensitivity and improve resistance to environmental stress cracking. They are used in the car industry for instrument panels, spoilers, and glove box flaps; in the electrical industry; and for household appliances like coffee machines, hair-driers, and flat iron handles. PC/ABS blends are marketed by Bayer, BASF, and DSM.

PC-polystyrene blends are used for similar purposes. They have a lower impact resistance than the unblended polymer but, in comparison with PC-ABS blends, are claimed to have better hydrolytic resistance, lower density, and higher heat deflection temperatures. They have been used for dishes for microwave ovens and for car headlamp reflectors.

PC-PBT blends with improved low temperature properties and petrol resistance are widely used for bumpers and front ends. They are marketed by GE Plastics.

PC-PET blends are somewhat similar. They offer improved chemical resistance against fats, greases, mild alkaline solutions, and a wide range of solvents. They are used in the chemical, food, and automotive industries and are marketed by Dow and GE Plastics.

Blends based on polyamides and polyimides

Block copolymers of PA with various polyethers have been developed by Monsanto to adapt PA to RIM, thus making PA behave like PU. However, it results in thicker wall parts; in the unnecessary need for lubrication; and in the needless use of toxic isocyanates. Work is understood to be still in the development stage. Possible markets include exterior car body components and appliance and business machine components.

PA-polyolefin alloys are tailored for resistance to automobile fuels, with low permeability and excellent burst strength. They are marketed by Monsanto and Atochem.

PA-modified EPDM rubber blends (the EPDM being treated with maleic anhydride) are the main example of the development of "supertough" nylons.

PA-PI compositions were developed to get round the intractability in the processing of unmodified PI. They are marketed as "Torlon", and are used both for compression moulding and injection moulding. They are very strong, stiff, creep resistant and have low water absorption. Uses include pumps, valves, gears, accessories for refrigeration plant, and electronic components. Low-friction materials can be made by blending this blend with graphite and PTFE.

Polybismaleimides combine polyimide with bismaleimide (itself the result of reaction between maleic anhydride and amines). Properties are broadly similar to polyimides and polyamide-imides. Unfilled polybismaleimides are used to make laminates, printed circuit boards and filament windings. Grades filled with glass fibre, asbestos, carbon fibre, molybdenum sulphide, graphite or PTFE are used in aerospace construction and in rocket and weapons technology. Specific uses include brake equipment, rings, gear wheels, friction bearings, and cam discs.

Polyester blends

PBT-PMMA blends, PET-PMMA blends, PBT-polyether-ester rubber blends, PBT-silicone/PC block copolymers and PBT-PET-PC blends all have enhanced strength. They are usually used in reinforced form. In addition to glass fibres other fillers such as glass beads, talc, mica, and carbon fibre are found.

PBT-polybutadiene rubber blends have improved toughness but have only moderate thermal aging qualities.

PBT-PMMA blends, on the other hand, have better heat-aging properties but are poorer in impact resistance.

PBT-PET blends, which to some extent combine the properties of the two polyesters, are marketed by GE Plastics.

Polyacetal blends

PMMA/PU alloys or blends show improvement in their toughness in two respects: higher strength under impact, and better elastic recovery.

Blends based on p-phenylene group derivatives

The PPS-PEK blend is marketed by Amoco, and PPO-PA blends are marketed by GE Plastics, BASF, and Asahi Glass.

PPO-high impact PS blends, which improve the impact properties of PPO in high performance grades, are marketed by GE Plastics, BASF, and Asahi Glass. PSU-PET blends are marketed by Amoco.

PSU-ABS blends (or Mindel blends) have lower heat resistance than unblended PSU, but are cheaper, easier to process and have higher notched impact strengths. Unreinforced blends are marketed by Amoco as Mindel A (a similar material being marketed by USS Chemicals as Arylon T); glass-fibre reinforced grades as Mindel B; and blends reinforced with other mineral fillers as Mindel M.

PSU-SAN blends, with similar properties, are marketed by Amoco as Ucardel. PPE/polystyrene and PPE/PA blends have been developed by BASF as a medium to high performance grades.

Fluorine-containing polymers

Tetrafluoroethylene-hexafluoropropylene copolymers, or Teflon FEP resins, were first developed by Du Pont and are now also produced by Daikin Kogyo (Japan) and Niitechim (USSR). They are mechanically similar to PTFE but have a greater impact strength, electrical insulation properties, and chemical inertness. It is exceptionally non-adhesive. Unlike PTFE, it can be injection moulded and extruded. Teflon is used for a variety of electrical and chemically resistant mouldings, and for corrosion resistant linings, coatings, flexible printed circuits, and wire insulation.

Hoechst markets Hostaflon TFB, a terpolymer of tetrafluoroethylene, hexafluoropropene, and vinylidenefluoride. It has similar properties.

Tetrafluoroethylene-ethylene copolymers (ETFE) are similar to Teflon but have a much higher abrasion resistance. It is marketed by Du Pont, with similar formulations being available from Hoechst and Asahi Glass. Applications are largely in the field of electrical applications, particularly high performance wire insulation.

PCTFE-ethylene copolymers have properties similar to PA, with low creep, good impact strength, and good chemical resistance and electrical resistance properties. They are resistant to burning. They are marketed by Allied Chemical.

Hexafluoroisobutylene-vinylidenefluoride copolymers, marketed by Allied Chemical, are similar to PTFE, although much less dense. They can be injection moulded and are less tough than PTFE.

Acrylic blends

Most blends of PMMA are directed at improving its toughness. Acrylonitrile-methyl methacrylate copolymer (Plexidur), marketed by Rohm and Haas), has good clarity, rigidity and surface hardness. It is most used as a glazing material for schools, sports halls, and vehicles.

PMMA-poly (butyl acrylate) blends (High-impact PMMA) have higher impact strength, but generally lower values for tensile strength and impact modulus than ABS, whose uses are similar. Commercial materials are marketed by ICI (Diakon MX), Rohm and Haas (Oroglas DR), and Rohm GmbH (Plex 8535-F). Other PMMA based moulding elastomer impact-modified grades are available from BASF.

Polyurethane blends

PU-POM grades are marketed by Hoechst. PU-acrylic blends have been produced, blending polyurethane intermediates with methyl methacrylate monomer and an unsaturated polyester resin to create an interpenetrating polymer network. Use of the acrylic monomer lowers cost and viscosity, while blends with 20% MMA + polyester have enhanced impact strength.

PU-styrene blends (PU cross-linked with PS) have high toughness, and high tensile and flexural strength. It is aimed at large mouldings, including automotive body panels, high speed speciality boats, and pultruded profiles. They are marketed by DSM.

Styrene copolymer blends

SAN-ABS blends are available to improve the impact resistance of SAN. SAN-chlorinated polyethylene blends result in ABS-type materials known as ACS. They have better flame retardancy, heat resistance, weatherability, and resistance to dust deposition than ABS, but have a poorer processing stability. SAN-olefin elastomer blends (olefin modified SAN) also give similar properties to ABS, but with better weathering properties.

ABS-alpha-methylstyrene alloys give high heat grades which have a lower melt viscosity at a similar cost. ABS-PVC blends are used to obtain fire retardancy in proportions of 20:80. They are tough but hard to process. In proportions of 10:90, they represent impact modified forms of unplasticized PVC. ABS materials have also been blended with plasticized PVC to give a crashpad sheet material.

ABS-PC blends are also fire retardant. They are easier to process than PVC blends but less tough. They provide possible alternatives to Noryl type polymers, and are marketed by Bayer and GE Plastics. ABS/PA blends are marketed by GE Plastics and Monsanto.

ABS-PSU blends are similar to ABS-PC blends, but have a higher heat distortion temperature and softening point. They are marketed by Amoco Chemical.

ABS-PMMA blends can produce a transparent ABS-like polymer. More complex terpolymers involving ABS and PMMA are also available, but are not widely used.

II.4 Flame-retardant, non-toxic filled polyolefins

Polyolefins occupy a leading position among the fast developing thermoplastic polymers. Their application possibilities in electronics, construction and the automotive industry are limited, however, by their inflammability. Recent developments show that the solution of the problem of fire resistance is a precondition for the competitivity of polyolefins with other materials in these applications.

The oldest and still most common way of producing flame-retardant polyolefins is by the addition of a combination of halogen (chlorine or bromine) containing hydrocarbons and antimony trioxide. During thermal decomposition of flame-retardant polyolefins of this type, toxic and corrosive smoke and vapours are evolved. In addition, the flame-retardant additives are effective at relatively low temperatures only; at higher temperatures they actually accelerate the burning process. The application of flame-retardant polyolefins of this type is thus limited and in some cases even not allowed.

A new solution of the problem brought the application of compounds containing either phosphorus or nitrogen, especially in combination with systems forming protective coke layers under the influence of excessive heat.

An alternative approach to the problem of flame-retardant polyolefins is the application of mineral fillers. They can play two roles in this case: they reduce the specific heat of combustion of the composite and, in some cases, they help to extinguish the flames by releasing non-combustible products of thermal decomposition. $Al(OH)_3$ is the best known filler with flame retardant properties: when heated to over 180° C, it dehydrates and evolves water vapour. Both the cooling effect of the endothermic reaction and the effect of dilution of oxygen and combustible gases by inert water vapours make the composite with high content of $Al(OH)_3$ self-extinguishing. However, $Al(OH)_3$ is not suitable for use in polyolefins, as its dehydrating temperature is lower than the processing temperature of polyolefins, especially of polypropylene.

A new filler, $Mg(OH)_2$ with a special morphology of particles has been developed by the KYOWA Corporation of Japan. This filler has a higher decomposition temperature and is suitable for use in polyolefins. Composites of polypropylene or polyethylene with high contents of $Mg(OH)_2$ are self-extinguishing (non-flammability V-0 according to UL-94 method) and exhibit good mechanical properties. In addition, the evolution of smoke and toxic and corrosive gases under the conditions of a fire is suppressed. The filler is also suitable for flame-retardant polyamide production.

A new group of fire-safe materials is represented by polymer composites with a very high content of inorganic phase (70% and more). These composites, called ceramoplastics, possess the advantages of both thermoplastic polymers and ceramics. In the case of fire, the inorganic phase forms a porous ceramic residue that prevents the fire from spreading further.

The chemical industry in the Czech and Slovak Federal Republic produces polyolefin composites with $CaCO_3$ or talc as filler, which successfully replace some more expensive engineering (ABS, polyamide) plastics in many applications.

In an effort to exploit the rich domestic reserves of pure magnesite and dolomite, an original technology has been developed which makes it possible to produce high-quality filler on the basis of $Mg(OH)_2$ with very good flame-retardant properties and good processibility. It is quite comparable with the KYOWA product. Polyolefin-$Mg(OH)_2$ composites are suitable for application in the electronics and the automotive industry (e.g., critically exposed parts in electronic appliances, automotive electrical components, etc.), and further applications are being developed.

Typical physical and mechanical properties of various types of flame-retardant polyolefins (both commercial and developmental products) are summarized in Tables 6-8.

Flame-retardant polyolefins with $Mg(OH)_2$ as an active flame-retardant filler have still not reached mass-production scale probably because of the need for a high degree of loading and the rather high cost of the product. It is believed, however, that the unique combination of mechanical properties of the composites and their fire behaviour will ensure a rapid expansion of their applications in the near future.

A new type of precipitated $CaCO_3$ can be obtained as a by-product by the processing of dolomite to $Mg(OH)_2$. The special morphology of the particles and the high purity of whiteness of this filler give the composites a combination of properties not attainable with talc or $CaCO_3$ of current types: high degree of stabilization, whiteness, toughness, stiffness and strength (Table 9). The composites can replace ABS copolymer in some cases but their main application potential is in fields where both good mechanical properties and a very pleasing appearance of the products are required.

Table 6. Properties of polyolefins filled with 60% w/w $Mg(OH)_2$

(informative values)

Polymer		PP Mosten				PE Liten
Property	Unit	55.212a/	58.412a/	52.692a/	52.511b/	FB 29
σ_k	MPa	20.2	20.1	22.6	18.1	24.6
a_k 23° C	kJ/m^2	4.7	2.2	1.6	NB	NB
E	GPa	4.5	4.7	4.8	3.1	3.3
MFI 21N	g/10 min	0.9	3.3	7.7	5.7	0.1
UL-94	(3.2 mm)	V-1	V-0	V-0	V-0	V-0

NB - no break.

MFI - Melt flow index.

a/ homopolymer.

b/ controlled rheology copolymer.

Table 7. Comparison of composites filled with 60% w/w $Mg(OH)_2$

(informative values)

Property	Unit	Magnesium hydroxide	
		VUANCH	KYOWA
σ_k	MPa	20.9	22.3
ε_r	%	7	7
a_k 23° C	kJ/m^2	3.0	2.5
E	GPa	4.4	4.5
Vic B 49N	° C	96	94
HDT B 0.45MPa	° C	125	119
MFI21N	g/10 min	5.4	4.9
UL-94	(3.2 mm)	V-0	V-0

Source: Country contribution.

Matrix- PP Mosten 58.412.

Table 8. Comparison of retarded polypropylene composites

(VUMCH informative values, injection moulded specimens)

Producer		Chisso Japan	Hilmont Italy	Hilmont Italy	Schulman F.R.G.	G.D.R.	CSFR	CSFR	Hilmont United States
Types		Chisso Polypro 2654	Moplen E1X94J	Moplen EPT94JR	Polyflam 371ND	Scolefin 2061	VUMCH PZ56Y60	VUMCH PZ56Y55	Pro-fax SD255
Filler		?	?	?	talc	talc	$Mg(OH)_2$	$Mg(OH)_2$	$Mg(OH)_2$
property	unit						a/	a/	b/
σ_k	MPa	29.9	27.6	13.9	20.0	28.3	20–22	21–23	20.0
a_k 23 °C	kJ/m^2	2.6	2.5	NB	3.8	2.3	2–4	3–6	NB
E	GPa	3.4	2.1	1.3	2.4	3.6	4.5–5.5	4.0–4.8	2.0
MFI 21N	g/10 min	4.0	8.9	3.7	15.4	1.5	3–5	5–7	5.0
UL-94	(3.2 mm)	V-0	V-0	V-0	V-0	V-0	V-0	V-1	V-0
LOI	% O_2	27	37		27	27	31	29	

Source: Country contribution.

NB - no break.

a/ matrix PP Mosten 58.412.

b/ values taken from preliminary data sheet.

Table 9. Comparison of composites filled with 40% w/w $CaCO_3$ and talc

(informative values)

Property	Unit	Filler content 40% w/w		
		Talc GT 40	$CaCO_3$ Durcal 2	$CaCO_3$ VUANCH
σ_k	MPa	31.4	24.5	27.2
ε_r	%	16	73	47
a_k 23° C	kJ/m^2	3.2	4.2	3.8
0° C		1.9	2.0	2.1
-20° C		1.5	1.8	1.8
E	GPa	4.2	2.9	3.7
Vic B 49N	° C	100	106	104
HDT B 0.45MPa	° C	136	127	131
MFI21N	g/10 min	3.1	2.4	3.1

Source: Country contribution.

Matrix - PP Mosten 58.412.

Legend

σ_k - yield strength (ISO 52A, DIN 53 455, ASTM D 638)

ε_r - elongation (ISO 327, DIN 53 455, ASTM D 638)

a_k - Charpy impact strength (ISO 179, DIN 53 453)

MFI - melt flow rate (ISO 1133, DIN 53 735, ASTM D 1238)

E - modulus of elasticity (ISO 178, ASTM D 790)

Vic - Vicat (ISO 306, DIN 53 460, ASTM 1535)

HDT - heat distortion temperature (ISO 75, DIN 53 461, ASTM D 1535)

UL-94 - non-flammability according to UL-94 (ANSI/UL 94-1979)

LOI - limited oxygen index (ASTM D 2863-70).

CHAPTER III. PLASTICS AND THE ENVIRONMENT

III.1 Introduction

Substances may damage the environment in a number of different ways:

- they may poison workers in the workplace,
- they may poison people who live in the neighbourhood of the factory producing them,
- they may adversely affect the biosphere of the world as a whole, and
- they may clutter up the environment with unsightly waste.

Furthermore, substances can be damaging:

- during manufacture,
- during processing,
- during use, or
- during destruction (either after use or accidentally, as in a fire).

Toxic effects on workers at the workplace occur during manufacture or processing. Toxic effects on people living in the neighbourhood generally occur during manufacture or destruction. Adverse effects on the biosphere may also occur either during manufacture or during destruction. The problems involved in waste *per se* occur at the time of disposal.

The above comments, on substances in general, relate to plastics in particular. As plastics are inert substances after manufacture and processing, they do not give rise to toxic or environmental problems during use. The only exception is when they are destroyed as a result of inappropriate use. This, however, is a particular kind of destruction.

Engineering plastics are affected by the same considerations as plastics in general. While specific plastics may have specific problems with regard to toxicity, there is nothing about engineering plastics in themselves that differentiates them from other kinds of plastics. There is one exception: the heterogeneity of engineering plastics creates problems of disposal after use different in kind from the problems involved in commodity plastics used for packaging.

The introductory discussion above leads to the conclusion that the environmental problems presented by engineering plastics may best be presented under the following four headings:

(a) Toxicity, both at the workplace and in the neighbourhood;

(b) Effects on the biosphere during manufacture or disposal;

(c) Problems of flammability;

(d) Problems of disposal.

These headings do not necessarily embrace all the environmental problems involved in the manufacture, processing and disposal of plastics in general, or engineering plastics in particular, but it can be argued that these problems are the most important ones.

It should be noted that all these areas are covered - and are increasingly being covered - by national and international regulations. It is not the aim of this study to cover all the relevant legislation.

III.2 Toxicity

A material may be considered toxic if it has an adverse effect on health. While some materials may have an immediate effect, their adverse effects (e.g. asbestos) may not emerge for many years. Some toxic materials can be purged from the body, others can accumulate until an eventual lethal dose may be present.

Toxic chemicals can enter the body in various ways, particularly by ingestion, inhalation, and skin absorption. All of these are relevant to plastics. People do not usually eat plastics or their constituents, but ingestion can take place if dust or fumes are allowed to contaminate food or drink. Similarly, inhalation can take place through the contamination of cigarettes. Skin absorption can lead to dermatitis. Certain halogenated aromatic materials such as formaldehyde and aliphatic amines have a tendency to cause dermatitis. Glass fibre can cause a different type of dermatitis through physical, rather than chemical, irritation.

Some chemicals have an almost universal effect on human beings, while others affect only a few. This could have a genetic element involved. A person who has worked with a given chemical for some years may become suddenly sensitized to it and, from then on, may be unable to stand the slightest trace of that material in the atmosphere. Formaldehyde, a chemical contained in PU foam, has apparently had this effect on people exposed to formaldehyde fumes from PU cavity insulation. The person may, as a result, become sensitized both to a specific chemical or to others closely related to it.

Chemicals can attack the body both externally and internally in many ways. Examples applicable to the plastic industry include lead salts, phenol, aromatic hydrocarbons, isocyanates and aromatic amines. Most west European countries have regulations aimed at minimizing the dangers presented by these chemicals, which are often used in trace quantities. Problems generally arise from the additives used, not from the polymer.

The following list shows some of the most important toxic substances which may be encountered in the manufacture and processing of engineering chemicals. Chemicals are only considered when they are used during the polymerization process, or thereafter. Chemicals used in monomer manufacture are ignored.

- Acrylonitrile: (used in the manufacture of ABS and SAN polymers) contains cyanide ions and is rated extremely toxic.

- Antimony salts: used as fire retardant additives, are poisonous when ingested (with similar effects to arsenic). They can also cause skin irritation.

- Asbestos: exposure can lead to pathological changes in the lungs. It was used as a reinforcing agent and also (in small quantities) as a catalyst. Realization of its very serious long-term toxic properties led to a ban (or at least stringent restrictions) on its use in most west European countries.

- Bromine compounds: are often added as fire retardant additives, particularly to ABS. While most concern is directed at their use as fire retardants, they do present some risk in manufacture. They depress the central nervous system.

- Cadmium: used as a colouring agent for some engineering plastics. It has adverse effects on the nerves and the blood. Pressure from Scandinavian Governments has led to its decreasing use in other west European countries.

- Carbon monoxide: one of the products of combustion of most plastics, has a very toxic effect by converting oxyhaemoglobin in the blood to carboxyhaemoglobin.

- Fibreglass: used for nearly all plastics as a reinforcing agent, it has a dermatitic effect.

- Fluorinated compounds: are formed during the sintering process used to make PTFE. In the form of fumes, they can severely irritate the lungs, and effective ventilation equipment is needed to remove them.

- Formaldehyde: used in the manufacture of acetal and PU resins, it can cause severe sensitization effects.

- Hydrochloric acid: in gaseous form, it is a product of combustion of chlorine-bearing plastics. It is a severe irritant.

- Isocyanates: used in the manufacture of PU, it presents similar hazards to acrylonitrile due to their cyanide content.

- Methyl methacrylate: (used in the manufacture of PMMA) has a skin irritant effect.

- Talc: used sometimes as a filler, it can, after prolonged exposure, lead to talcosis, a condition of the lungs.

On the whole, with the possible exception of PU, engineering polymers do not present exceptional toxic hazards. PVC and epoxy resins generally present more severe and more intractable problems. It should be stressed, however, that any kind of dust or solvent may present a health problem.

III.3 Effects on the biosphere

Many chemicals are thought to have an adverse effect on the biosphere, but the most important ones as far as plastics are concerned are chlorinated fluorocarbons (CFCs), which are used in the manufacture of foamed polyurethane (a slightly different form of CFC is used on extruded polystyrene foams).

It is thought that chlorinated fluorocarbons, halons, methane, nitrous oxide, and carbon monoxide, have an important effect in reducing stratospheric ozone levels, particularly in the Antarctic. This might lead, in turn, to an increase in hard UV radiation reaching ground level, with possible detrimental effects on health (e.g. an increase in the level of skin cancers, and possible genetic effects).

Although the plastics industry is one of the major users of CFCs, they are used in other industries, such as in the production of refrigerators.

In accordance with the Montreal Protocol, an international agreement was reached to discontinue the production and use of CFCs by the year 2003. Subsequently, EEC countries agreed among themselves to backdate this by five years. Du Pont had announced earlier its intention to cease production of CFCs by the end of the century.

In the past few years a number of systems have been put on the market to replace CFCs. Substitutes include n-pentane, and hydrogenated fluorocarbon 22. Proprietary systems are also available, for example Reticel's LBL2, which is claimed to have an ozone depletion potential (ODP) value of nearly zero (0.003, as against 1.00 for CFC11), and a very short atmospheric lifetime.

It is evident that, in western Europe and the United States, replacement of CFCs by the plastics industry has proceeded well ahead of schedule. The European Isocyanate Producers Association has calculated that the European polyurethane industry can expect to achieve a 50-60% reduction in the use of CFC11 by the end of 1993. Further progress is likely to depend on the use of hydrofluoroalkane blowing agents, subject to satisfactory toxicity testing.

Requirements for the new formulations have proved to be significantly less than the CFCs which they replace.

However, problems with CFCs do not only occur during manufacture. In some PU foams, particularly rigid foams, CFCs are retained within the foam and may be released when it is disposed of. This is a problem which will be with the plastics industry for many years as CFC-containing foams are gradually disposed of. Emission of CFCs during disposal of old foams can only be avoided if this is done by landfill or by destruction in a closed circuit system. It is important that they should not be disposed of by burning with emission of the products of combustion to the atmosphere.

III.4 Problems of flammability

In this section, a broad view is taken of the effects of fire on the environment. Fire not only affects the environment by destroying buildings, but it can also involve the emission of lethal or harmful products of combustion.

Many plastic materials burn fairly readily. Engineering plastics are on the whole less prone to do so, but flammability is a problem with some which are widely used, in particular PU foam, POM, ABS and, to a rather lesser degree, nylon. PET and PBT are also inflammable to some extent, but most grades of these polymers are filled with glass fibre, which reduce fire risks considerably.

It is important that in uses such as building, furniture and transport, plastic products should have an adequate degree of fire resistance. This has led to the increasing use of fire retardants.

A flame retardant should inhibit or even suppress the combustion process. Depending on their nature, flame retardants can act chemically and/or physically in the solid, liquid or gas state. They interfere with combustion during a particular phase of this process, e.g. during heating, decomposition, ignition or flame spread.

There are several ways in which the combustion process can be retarded by physical action:

(a) by cooling: endothermic processes triggered by additives cool the substrate to a temperature below that required for sustaining combustion. Aluminium trihydrate operates in this way.

(b) by formation of a protective layer: the condensed combustible layer can be shielded from the gaseous phase by a solid or gaseous protective layer. The condensed phase is thus cooled, the oxygen necessary for the combustion process is excluded and heat transfer is impeded. Phosphorus compounds operate in this way.

(c) by dilution: the incorporation of inert substances and additives, which evolve inert gases on decomposition, dilutes the fuel in the solid and gaseous phases so that the lower ignition limit of the gas mixture is not exceeded. Aluminium trihydrate works in this way.

The most important chemical reactions interfering with the combustion process take place in the solid and gas phases:

(a) by reaction in the gas phase: the radical mechanism of the combustion process which takes place in the gas phase is interrupted by the flame retardant. The exothermic processes are thus stopped, the system cools down, and the supply of flammable gases is reduced and is eventually completely suppressed. Halogenated flame retardants operate in this way. Bromine and chlorine are the bases of most halogenated flame retardants, bromine being rather more effective.

(b) by reaction in the solid phase: two types of reaction can take place:

Breakdown of the polymer can be accelerated by the flame retardant causing the polymer to flow and withdraw from flame. Peroxides operate in this way.

A layer of carbon can be made to form on the surface. This may occur through the dehydrating action of the polymer generating the formation of double bonds in the polymer. These form the carbonaceous layer by cyclizing and cross-linking. Phosphorus compounds operate in this way.

A distinction is made between reactive and additive flame retardants. Combinations of retardants may produce a synergistic effect.

Reactive flame retardants are built chemically into the polymer molecule. They are used for the most part in thermosetting resins and polyurethanes. They are usually either highly halogenated or are phosphorus compounds.

Additive flame retardants are incorporated into the plastic before, during or (usually) after polymerization. They are used especially in thermoplastics.

Combinations of retardants can produce an additive or synergistic effect. When used alone, synergistic additives show little effectiveness, but the synergistic effect occurs in combination with other additives. Antimony oxide, which is synergistic with phosphorus and halogen compounds, are widely used. Others include titanium dioxide, zinc oxide, and molybdenic oxide. Zinc borate has also been used.

Aluminium trioxide is generally used for plastics formulations with processing temperatures up to about 190° C. To obtain good flame retardance, large quantities of additive are required.

Inorganic flame retardants like zinc borate and magnesium hydroxide can be used for higher processing temperatures. Here too, large quantities need to be used.

Brominated flame retardants are commonly used for thermoplastics. Decabromo diphenyl ether (Deca) is used in most formulations of engineering plastics. Octabromodiphenyl ether (Octa) is used mainly in ABS.

For engineering plastics like PBT, PET, PC and PA various brominated flame retardants like ethylene bis tetrabromophthalamide, brominated polystyrene, and derivatives of tetrabromobisphenol A are commonly used.

Aliphatic and cycloaliphatic chlorinated compounds are applied in PA. Red phosphorus is an effective flame retardant for PA. It has the disadvantage, for some uses, of making the products dark in colour.

Smoke

Fire often involves the production of smoke. The amount produced depends on the source of ignition, oxygen availability and the properties of the material which is burning. Smoke endangers life, particularly in the initial stages of a fire. People in a fire can lose vision and/or become disorientated. In these circumstances it has been shown that they tend to give up trying to escape and stand still, to be asphyxiated or overcome by other toxic gases like carbon monoxide. Alternatively, they may be burned to death. Thus while the direct effects of smoke are not necessarily lethal, the indirect effect can be very damaging.

This has led to work being carried out on the use of smoke suppressants to decrease the smoke generating potentiality of a number of plastics, including PU, which is important in view of its use in furniture and many transport applications.

In many cases, there is a trade-off between the use of flame retardants and the use of smoke suppressants. Fire retardants which are effective in the gas phase for thermoplastics can cause intense smoke development, but the addition of smoke suppressants chemically active in the gas phase can interfere with the flame retardant effect. Physically acting smoke suppressants which act simultaneously as flame retardants are of limited efficacy. Use of fillers may affect the mechanical properties of a plastic in undesirable ways. There is room for much development work in this area, as the problems have by no means been solved.

Emission of poisonous gases during combustion

The main toxic agent produced by plastics in a fire is carbon monoxide. Other toxic agents which may be present include hydrogen cyanide and hydrogen chloride, which are likely to be emitted by any substance containing nitrogen and chlorine, respectively.

Concentrations of these gases in the initial stages of a fire are likely to be low, and people are unlikely to stay around long enough to suffer ill effects from them, provided means of escape are at hand and they are not disabled by smoke. In certain circumstances these conditions may not be obtained - for example in fires underground. Halogen-containing materials tend to be ruled out completely (as in London Underground specifications) in these situations.

Generally speaking, smoke from fires has received more attention than toxicity, but in the longer run this situation may change. Problems of toxicity seem to be attracting more attention throughout the European continent, e.g. from the French railway system SNCF.

In Germany, concern has been expressed that the polybrominated diphenyl ethers pentabromo- (Penta), octabromo- (Octa), and decabromo diphenyl ether (Deca), which are widely used as flame retardants, may be the precursors of potentially dangerous polybrominated dibenzodioxins (PBDDs) and dibenzofurans (PBDFs). There is some (but conflicting) experimental evidence for this. In any case, additive manufacturers may well try to find other formulations.

III.5 Disposal

Possibly the most intractable problem relating to plastics and the environment is that of their disposal. Even apart from the disfiguring effect of plastic litter, the disposal of plastics presents a pressing problem.

Plastics have hitherto been disposed of by landfill or by burning. Both courses are objectionable from an environmental point of view, and also few people are willing to put up with their presence in their neighbourhood landfill sites or incineration facilities. Landfill sites are becoming very scarce. One estimate is that there are less than 20 years' landfill capacity left in western Europe. Indeed, there are press reports that waste contractors and local authorities are now seeking sites in eastern Europe, thus exporting a problem which they cannot solve.

Recycling is becoming increasingly attractive, because:

(a) it obviates the need for landfill sites;

(b) it avoids the need for incineration plants where toxic or noxious fumes may be emitted; and

(c) it husbands resources by lessening the demand for primary materials.

Recycling packaging waste has been practised for some years, and is encouraged both by the EC Commission and by public authorities. Draft directives are under discussion by the former (e.g. the EC Council Directive 339/85 on one-way beverage packs). Many public authorities throughout Europe are encouraging plastic waste recycling through the institution of bottle banks and the like. Other forms of packaging waste can easily be collected at retail stores or at other points in the distributive chain.

Recycling waste engineering plastics is much harder. Relatively few plastic materials are used for packaging, and in many (not all) cases it is fairly easy to distinguish PVC from PET from polyolefins. In the case of engineering plastics, many more different formulations are in common use, and distinguishing one from another is by no means easy. The most pressing problem arises from automobile waste, where up to 20 different engineering plastics may be encountered in a single model. Further discussion in this section will relate largely to automobile waste.

The EC Commission has stated its intention to produce a draft Directive on automobile waste, and the details of this are currently under discussion. The declared aim is that a high proportion of plastic automobile waste should be recycled by the end of the century.

At present, recycling plastic automobile waste presents considerable difficulty, for a number of reasons:

(a) Most dismantling of cars is at present carried out by scrap dealers who are oriented toward iron and steel, not plastics.

(b) Much of the plastic in a car is not easily reached. Currently several car firms are carrying out experimental dismantling to see how removal may be facilitated.

(c) Even if the plastic is removed from the car, the problem remains of identifying it. Scrapyard workers are not likely to be plastics technicians.

(d) The plastics are not likely to be compatible with one another, and in any case there is little market for mixed waste plastic.

(e) Many engineering plastics are heavily loaded with glass fibre, and other additives may be used which may not be compatible with each other in their action.

Several approaches are being made to this problem. In solving it, due attention must be paid to the fact that the subject is engineering plastics, not packaging. Thus solutions, based on a flow of identical material, must be disregarded (since plastic automobile waste is very heterogeneous) and biodegradability is infeasible (biodegradable cars would have little use or attraction). Furthermore, developing a market for mixed automobile waste is not promising. Possible solutions include:

(a) Developing automatic ways of sorting (an example exists of a method developed of sorting PVC, PET and polyolefin bottles). The PVC bottles are isolated by electrical means which depend on the identification of the presence of chlorine atoms. The remaining bottles are ground down and PET is separated from polyolefins by a flotation process in which the PET granules sink and the polyolefin granules stay at the top. Automatic ways of separating out engineering plastics would be complex, but are at least conceptually possible.

(b) A similar separation can be achieved by immersing mixed waste in a series of solvents which remove one plastic after another in a kind of cascade. This approach is at an early experimental stage so far.

(c) Another approach is to abandon the idea of separation and have all mixed plastic automobile waste undergo pyrolysis. The output of this process could be used as a feedstock for the manufacture of plastics.

While none of these approaches is viable so far - at least as regards to engineering plastics - it does seem likely that in 10 or 20 years' time, disposal will occur by one or more of these methods.

As stated above, the problems in this area are not at all tractable, and there is a need for serious research to be developed in this area.

CHAPTER IV. MARKET FOR ENGINEERING PLASTICS IN THE MARKET-ECONOMY COUNTRIES

IV.1 Acrylonitrile butadine styrene (ABS) copolymers

ABS copolymers occupy an important place in the range of engineering plastics and are among the most promising copolymers. The mean annual growth rates for ABS copolymer production will remain fairly high in the future, surpassing those for polystyrene, as can be seen from the data provided below (%). 1/ 2/

Polymer	1983/1989	1989/1995 (forecast)	1995/2005 (forecast)
World			
ABS copolymers (including SAN copolymers)	7.0	4.3	3.3
Polystyrene	7.1	3.5	2.9
Western Europe			
ABS copolymers (including SAN copolymers)	7.0	4.0	-
Polystyrene	7.0	3.0	-

The United States holds first place for total production of ABS copolymers, but it is in Japan that their production is increasing most rapidly. In the period 1981-1989, the mean annual growth rate of production was 3.45% in the United States and 8.75% in Japan. ABS copolymer production in the United States showed a decrease of about 7% in 1990 compared with the previous year. 3/-6/

However, it should be noted that, at present, quantitative growth in the production of polymers (especially high-tonnage polymers) abroad is not the decisive factor in evaluating the status of their production. The main trend in recent years in the synthetic resins and plastics industry has been a shift from quantitative to qualitative growth in the production and consumption of general-purpose plastics, their transformation into plastics for technical and later for engineering uses. The most suitable materials from this point of view are the ABS copolymers. An outstanding feature of these copolymers is that they combine high strength characteristics with relatively low cost. A wide range of ABS copolymers (more than 15 types) are produced abroad on an industrial scale, a considerable proportion of them being special-purpose materials such as heat-resistant, weather-resistant, metallized, glass-fibre-reinforced and self-extinguishing grades, alloys and blends, and structured foam plastics.

ABS copolymer consumption worldwide totalled 1,610,000 tonnes in 1985 and was expected to be 1,750,000 tonnes in 1990. It is forecasted to rise to 2,370,000 tonnes by 1996. 7/-9/ Japan occupies first place for consumption per head of these copolymers (4.1 kg in 1990).

Experts believe that growth in ABS copolymer consumption will depend mainly on the development of markets such as expensive large-dimension instruments, portable tape recorders and televisions, telecommunications and office automation equipment. 12/-14/

It is estimated by foreign experts that $US 1.3 billion worth of high-tonnage engineering thermoplastics and $US 2.3 billion worth of high-heat-resistant thermoplastics were sold worldwide in 1988. Consumption of these materials is expected to grow at mean annual rates of 7-8% and 10-12% respectively up to the mid-1990s. 15/ 16/ Specialists at General Electric Plastics forecast that production of high-tonnage engineering thermoplastics worldwide will exceed 2.5 million tonnes in 1992 (1.9 million tonnes in 1988). 17/-19/ The most important materials in this group in terms of production and consumption are the polyamides, for which consumption growth is comparable with production growth and is estimated at 6% for the period from 1988 to 1993. Polyamide consumption in North America will increase from 272,000 tonnes in 1988 to 354,000 tonnes in 1993. 19/ 20/ In western Europe, according to a forecast by Frost and Sullivan, polyamide sales will reach 285,000 tonnes in 1993. In Japan polyamide production is expected to rise to 188,000 tonnes by 1995. 21/ 22/ The most promising polymers in the high-tonnage engineering thermoplastics group are the polycarbonates. It is forecasted that the mean annual consumption growth for polycarbonates will be 8-10% in 1989-1993 and that their consumption in the market-economy countries will increase to about 386,000 tonnes by 1993, exceeding the forecasted consumption for polyamides.

Up to the mid-1990s the most important applications for high-tonnage engineering thermoplastics will continue to be in transport and electrical engineering, electronics and office equipment.

The American firm Kline and Company estimates that the world market for high-heat-resistant thermoplastics will reach $US 5 billion by 1998. 18/ However, despite the large forecasted sales and significant rates of growth in the consumption of these materials, it is expected that they will remain low-tonnage products for special uses and that consumption in the mid-1990s will not be above 100,000 tonnes.

North America will retain its importance as the major supplier and consumer of high-heat-resistant engineering thermoplastics. The most intensive growth for these materials will occur in the Asian countries. 23/

The leading applications for high-quality thermoplastics will continue to be in the aerospace and electronics industries. They will be used increasingly in the automotive and mechanical engineering industries as their cost decreases. 18/

One of the most promising polymers in this group is polyphenylene sulphide. Specialists at Hoechst Celanese (United States) forecast a 15% annual increase in the consumption of this material up to 1995 and a 10% annual increase from then until the end of the century. 24/ Polyphenylene sulphide consumption will almost double in the United States between 1990 and 1995, from 8.6 to 15.9 thousand tonnes, and will reach 37.3 thousand tonnes in the market-economy countries by 1995. 23/ 25/

Other promising engineering plastics include liquid crystal polymers, whose consumption at present is low (450 tonnes/year in 1988-1989), but the American company R.M. Kossoff and Assoc. estimates that it may reach 11-14,000 tonnes/year provided that prices drop by half. In the past few years manufacturers have developed several new LCP grades costing less than the first-generation materials. Their introduction onto the market is expected to broaden considerably the applications of liquid crystal polymers in fields such as mechanical engineering and medicine. 26/

IV.2 Polyurethanes and other engineering plastics

Polyurethanes are used widely in many branches of the economy because of the great variety of materials which can be made from them. In most developed countries, foam plastics account for about 80% of total polyurethane consumption, but their share has been gradually declining as a result of the more rapidly increasing use of other types of products. Total consumption of polyurethanes roughly tripled in the market-economy countries between 1970 and 1985 to reach about 3.5 million tonnes. In the United States, their share in total plastics consumption in 1989 was 5.6%. In western Europe polyurethane consumption was 1,422,000 tonnes in 1984 and 1,508,000 tonnes in 1989. In Japan 289,000 tonnes of polyurethane foam was produced in 1988 and 308,000 tonnes in 1989. 27/ 28/ It is expected that demand for polyurethanes will rise over the next 10-20 years. According to a forecast by the firm Business Communications (United States), polyurethane foam consumption in the United States will grow at a mean annual rate of 4.3% in 1990-1995, reaching about 1.7 million tonnes in 1995. 29/ There has been a considerable increase in the share of reinforced polyurethanes made by reaction injection moulding. In the United States, this share had been only 2% of all RIM polyurethanes in 1980, but reached 40% in 1984. Products made by reaction injection moulding are used in the automotive industry and have good application prospects. 30/ 31/

Fluoroplastics belong in the category of special-purpose plastics. Consumption is low (48,000 tonnes in 1987, of which 31,000 tonnes of polytetrafluoroethylene, 7,000 tonnes of polyvinylidene fluoride and 10,000 tonnes of other fluoroplastics). Atochem (France) forecasts that fluoroplastics consumption in the market-economy countries will rise to 85,000 tonnes/year by 1995, with the share of polytetrafluoroethylene falling to 53% and that of polyvinylidene fluoride increasing to 21%. Significant growth in the demand for fluoroplastics is hampered by their high cost and difficulty of processing. 32/-35/

The domestic prices for some types of engineering plastics in the United States (August 1990) and Japan (January 1991) are given below: 36/ 37/

United States	$US/kg
ABS copolymers	
average impact strength	1.98-2.10
high impact strength	2.10-2.31
transparent	2.30-2.58
fire-resistant	2.71-2.87
Polyacetals	
homopolymer	2.76-2.87
copolymer	2.76-2.87
Polyamides	
6	2.69-2.79
66	3.02-3.04
69	5.51-6.08
610	6.30-6.90
612	6.30-7.21
11	7.25-7.52
12	7.01-7.52
Polyethylene terephthalate, modified	
containing 30% glass fibre	2.65-3.02
containing 55% glass fibre	2.98-3.24
Polybutylene terephthalate	2.98-3.02
Resins based on polyphenylene ether/polyphenylene oxide	
for pressure moulding	2.87-4.41
for extrusion	3.31-4.63
Polycarbonates	
for pressure moulding	3.04-3.20
for extrusion	3.20-3.42
for blow moulding	3.09-3.37
Thermoplastic polyurethanes	3.53-5.51
Polyarylates	4.41-6.17

United States	$US/kg
Polyphenylene sulphide	
containing 40% glass fibre	6.90-7.28
containing 20% glass fibre and 35% filler	3.46
containing 35% glass fibre and 35% filler	4.34
Polysulphone	8.64
Polyaryl sulphone	9.70
Polyether sulphone	9.70
Polyimide ether	10.25
Polyimide amides	
modified	38.58
unmodified	41.89
Liquid crystal polymers	
for pressure moulding	
with mineral filler	15.21-22.82
with glass-fibre filler	15.76-23.48
with carbon filler	37.48-44.10
for extrusion, unfilled	26.46-48.50
Fluoropolymers	
polytetrafluoroethylene	12.57-14.11
fluorinated ethylene propylene	20.40-24.03
ethylene and chlortrifluoroethylene copolymer	24.2
ethylene and tetrafluoroethylene copolymer	27.12-33.18
perfluoroalkoxylated copolymer	40.30-44.20
polychlortrifluoroethylene	66.14
Polyetherether ketone	50.70
Polyether ketone	65.04

Japan	Y/kg
ABS copolymers, impact strong	430-450
Polymethyl methacrylate	480-500
Polyacetals for pressure moulding	650-700
Polyamide 6	690-730
Polybutylene terephthalate	705-755
Polycarbonates	900-1000

As can be seen from the above data, the prices for engineering thermoplastics are fairly high. Even the cheapest of them - ABS copolymers - are more than twice as expensive as polyethylene and polypropylene, and almost three times dearer than polyvinyl chloride. The price difference between general-purpose thermoplastics and the highest-tonnage engineering thermoplastics (polyamides, polycarbonates, polyphenylene oxide) is $US 2-3 per kilogram. The prices for many types of high-heat-resistant engineering thermoplastics (modified polyimides, polyether ketones, liquid crystal polymers) are so high that they are used in small quantities and mainly in fields where there is no price competition.

The range of plastics manufactured in the leading market-economy countries is very wide. Existing modification techniques (copolymerization, grafting, blending, filling) provide great scope for altering the properties of plastics. Consequently, the most successful materials on the market are those which combine the requisite properties and fairly high performance characteristics with low cost. One of the major challenges, therefore, is to bring down the cost of engineering plastics. The prospects for their increased consumption depend significantly on how far this is achieved.

CONCLUSIONS

The list of polymer-based materials used in engineering and technical applications comprises 21 polymers and copolymers, including the so-called "super-engineering" plastics: polyamides, polyether sulphones, polyetherether ketones and liquid crystal polymers.

Growth in the production and consumption of engineering plastics is concentrated in the countries of western Europe, the United States and Japan, and is characterized by high rates of production growth (from 5 to 12% annually, or up to 80-85% for RIM polyurethanes). This trend is expected to continue at least until the year 2000, except for liquid crystal polymers, in view of their high cost.

New types of base polymers for engineering plastics are not expected within the next 10 years and consequently the highest growth rates will occur in the production and consumption of composite (reinforced) materials made from existing base polymers and blends or alloys of them.

Companies in the United States, Japan and Germany are the leaders in the production and consumption of engineering plastics, especially super-engineering plastics, mainly in view of the high R & D and capital intensity of plants producing these materials. The domestic market demand in these countries also contributes significantly to increasing production, since advanced mechanical engineering, electrical and electronics goods cannot be manufactured without these materials.

Notes

1/ Promt, 1988, 81, N. 4, 47.

2/ European Plastic News, 1991, 18, N. 1/2, 21.

3/ Chemical Market Reporter, 1990, 237, N. 7, 25.

4/ Plastic World, 1991, 49, N. 1, 3.

5/ Promt, 1989, 81, N. 4, 47.

6/ Japan Chemical Week, 1990, 31, N. 1552, 4.

7/ Promt, 1988, 80, N. 2, 49.

8/ European Chemical News, 1990, 54, N. 1417, 20.

9/ European Plastic News, 1985, 12, 5, 2.

10/ European Chemical News, 1989, 53, N. 1393, 23.

11/ Chemical Week, 1989, 145, N. 7, 18.

12/ European Plastic News, 1989, 16, N. 10, 36.

13/ Modern Plastics International, 1988, 18, N. 8, 46.

14/ Polymer News, 1989, 14, N. 4, 116.

15/ Science et technologie, 1990, N. 24, 52-53.

16/ Chemical Business, 1989, 11, N. 4, 44-46.

17/ Chemical Market Reporter, 1988, 233, N. 26, 9, 16.

18/ Chemical Market Reporter, 1989, 236, N. 24, 25, 28.

19/ Chemical Week, 1989, 144, N. 19, 16, 17.

20/ Trends in End-Use Markets for Plastics, 1990, N. 0109.

21/ Polymer News, 1990, 15, N. 2, 5.

22/ Trends in End-Use Markets for Plastics, 1990, N. 1570.

23/ Plastic Age, 1987, 33, N. 10, 128-132.

24/ Chemical Market Reporter, 1990, 238, N. 10, 5.

25/ Plastic World, 1990, 48, N. 11, 16.

26/ *Modern Plastics International*, 1989, 19, N. 10, 16.

27/ *Modern Plastics International*, 1990, 20, N. 1, 41-43.

28/ *Materielle Plastique et Elastiomère*, 1985, N. 11, 584-585.

29/ *Trends in End-Use Markets for Plastics*, 1990, N. 1794.

30/ *Reinforced Plastics*, 1981, 25, N. 3, 80.

31/ *Modern Plastics International*, 1986, 16, N. 1, 9-10.

32/ *Kunststoffe*, 1987, 77, N. 10, 49-51.

33/ *Chemical Week*, 1986, 139, N. 16, 21-24.

34/ *Modern Plastics*, 1988, 65, N. 9, 60-64.

35/ *Plastic Technologies*, 1989, 35, N. 1, 107-111.

36/ *Plastic Technologies*, 1990, 36, N. 9, 125-127.

37/ *Japan Chemical Week*, 1991, 32, N. 1606, 7.

ANNEXES

Annex I

LIST OF FIRMS ON ENGINEERING PLASTICS

Airtech Chemical Corporation
(USS Chemicals)
Thermoplastic Polymers Division
600 Grant St.
Pittsburg
PA 15230
United States of America
Tel (412) 433 7858
Fax (412) 433 7932

AKZO Plastics BV
Postbus 60
Velperweg 76
NL 6800
AB Arnhem
The Netherlands
Tel (085) 664422
Fax (085) 663434

Altulor
Tour Gan
18 Place de l'Iris Cedex 13
92082 Paris la Defense 2
France
(This is a subsidiary of CdF Chemie and hence of Atochem)
Tel (01) 4776 4117
Tlx 610804

Amoco Chemical (Europe) SA
15 rue Rothschild
1211 Geneva 21
Switzerland
Tel (022) 731 02 81
Tlx 845 422 787

Asahi Glass Co. Ltd.
1-2 Marunouchi 2-Chome
Chiyoda-Ku
Tokyo 100
Japan
Tel (03) 218 5555
Fax (03) 287 0772

Atochem SA
4 Cours Michelet
La Defense 10 - Cedex 42
92091 Paris la Defense
France
Tel (01) 49 008080
Fax (01) 47 738122

BASF AG
D-6700
Ludwigshafen AM Rhein
Germany
Tel (0621) 6042141
Fax (0621) 6041849

Bayer AG Polymer Division D-5090 Leverkusen Germany	Tel (0214) 301 Fax (0214) 308923
Dart Industries Ltd. Tupperware House 130 College Road Harrow HA1 1BQ Middlesex England	Tel (081) 861 1819 Fax (081) 861 0862
Dow Europe SA Bachtobelstrasse 3 8810 Horgen Switzerland	Tel (01) 728 2111 Fax (01) 728 2935
DSM-Marketing Centre Chemicals and Polymers Technical Department - Polymers Postbus 604 6160 AP Geleen The Netherlands	Tel (046) 69222 Fax (046) 63853
Du Pont c/o Nemours International SA Engineering Polymers CH-1211 Geneva 24 Switzerland	Tel (022) 378111 Fax (022) 433822
Eastman Chemical Products PO Box 431 Kingsport, TN 37662 United States of America	Tel (800) 327 8626 Fax (615) 229 2000
EMS-Chemie AG CH-7013 Domat/EMS Switzerland	Tel (081) 366111 Fax (081) 367401
Enichem Technoresine Spa Via Medici del Vascello 40 20138 Milan Italy	Tel (02) 5201 Fax (02) 52039590
Fudow Chemical Co. 4-11-26 Nishi-Rokugo Ota-Ku Tokyo 144 Japan	Tel (03) 737 0611

Address	Contact
General Electric Plastics Europe Plastics Laan-1 PO Box 117 4600 AC Bergen-Op-Zoom The Netherlands	Tel (01640) 32911 Fax (01640) 43949
Hoechst AG Postfach 803020 D-6230 Frankfurt-am-Main Germany	Tel (069) 3057160 Fax (069) 303 665/66
Huls AG PO Box 1320 D-4370 Marl Germany	Tel (02365) 491 Fax (02365) 49 2000
ICI Advanced Materials PO Box 6 Shire Park, Bessemer Road Welwyn Garden City AL7 1HD Hertfordshire England	Tel (0707) 323400 Fax (0707) 335556
Idemitsu International Europe Ltd. Economist Building 25 St. James St. London SW1A 1HA England	Tel (071) 839 554/8 Tlx 919638
Jonylon Ltd. Ramsey Works Fields New Road Chadderton Oldham Lancs England	Tel (061) 652 4411 Fax (061) 627 0102
Kureha Chemical Industry Co. Ltd. 9-11 Nikonhashi Horidome-Cho 1-Chome, Chuo-Ku Tokyo 103 Japan	Tel (03) 662 9611 Fax (03) 661 1277
Laporte Industries Ltd. PO Box Kingsway Luton LU4 8EW Bedfordshire England	Tel (0582) 21212 Fax (0582) 31818

Company	Contact
Mitsubishi Plastics Industries 2-5-2 Marumouchi Chiyoda-Ku Tokyo 100 Japan	Tel (03) 282 4111
Monsanto Europe SA Avenue de Terwren 270-272 B-1150 Brussels Belgium	Tel (02) 761 4111 Fax (02) 761 4040
Occidental Petroleum Corporation Durez Division (formerley Hooker Chemical and Plastics Corporation) 673 Walck Road PO Box 535 North Tonawanda New York 14120 United States of America	Tel (716) 696 6000 Fax (716) 696 6260
Pennwalt Plastics Ltd. Cherwell House St. Clements Oxford OX4 1BD England	Tel (0865) 726 961 Fax (0865) 250 321
Phillips Petroleum Chemicals NV/SA Brusselsesteenweg 355 B-1900 Overijse Belgium	Tel (02) 689 1211 Fax (02) 689 1472
Policarbonatos de Brasil SA R. mar Vermelho 474 CEP 03570 (Parque Savoy City) Brazil	Tel 422 2166 Tlx (011) 31995
Resart-IHM AG Gassnerallee 40 D-6500 Mainz 1 Germany	Tel (06131) 63 10 Fax (06131) 6311 42
Rohm GMBH D-6100 Darmstadt 1 Kirschenallee 45 Postfach 42 42	Tel (06151) 1801 Fax (06151) 184468

Rohm and Hass Company
European Operations
Chesterfield House
15-19 Bloomsbury Way
London WC1A 2TD
England

Tel (071) 242 4455
Fax (071) 404 4126

Solvay and Cie SA
Direction Centrale Plastiques
Rue du Prince Albert 33
1050 Brussels
Belgium

Tel (02) 5166111
Fax (02) 5166617

Teijin Chemical Ltd.
Daiwa Bank Toranomon Building
6-21 Nishi-Shinbashi 1-Chome
Minato-Ku
Tokyo
Japan

Tel (03) 506 4778
Fax (03) 506 4760

Annex II

DATA RECEIVED IN REPLY TO THE ECE QUESTIONNAIRE ON PRODUCTION IMPORTS, EXPORTS AND CONSUMPTION OF ENGINEERING PLASTICS

BELGIUM

Production, import and export of engineering plastics

(in tonnes)

Plastic	Production	Imports	Exports
	1988	1987	1987
Polycarbonate	23 000	n.a.	n.a.
Polyamide and polyimide	n.a.	23 113	35 608
Polyethyleneterephthalate	33 000	n.a.	n.a.
Polyurethane	10 000	27 021	59 482
Acrylonitrile-butadiene-styrene	100 000	n.a.	n.a.
Total engineering plastics	n.a.	n.a.	n.a.
Total of all plastics	2 561 000	1 508 000	2 830 000

BULGARIA

Consumption of engineering plastics in various sectors of national economies

(in tonnes)

Plastics	Engineering			Electronics			Automotive industry			Construction		
	1985	1988	1989	1985	1988	1989	1985	1988	1989	1985	1988	1989
1. Total engineering plastics	3 627	4 253	4 618	1 879	2 150	2 343	960	1 012	1 040	3 255	3 978	4 000
of which:												
- Polycarbonate(s)	180	210	215	250	300	340	10	10	10	10	10	15
- Polyamide	2 100	2 500	2 800	1 050	1 100	1 100	80	80	80	20	25	25
- Polyformaldehyde	40	65	75	65	130	170	-	2	5	5	8	10
- Polybutyleneterephthalate	-	1	2	2	8	11	-	-	-	-	-	-
- Polyethyleneterephthalate	4	6	8	5	11	20	-	-	-	-	-	-
- Polychlortrifluoroethylene	2	6	6	-	1	1	-	-	-	-	-	-
- Polytetrafluoroethylene	70	85	90	2	3	3	-	-	-	-	-	-
- Polysulphone	-	7	8	1	2	2	-	-	-	-	-	-
- Polyphenyleneoxide	1	3	4	4	5	6	-	-	-	-	-	-
- Polymethylmethacrylate	710	720	750	350	380	460	20	20	25	420	430	450
- Polyurethane	520	650	660	150	210	230	850	900	920	2 800	3 500	3 500
2. Composites												
Glass-fibre-plastics	220	2 250	2 250	300	310	310	20	40	40	420	350	350
3. Blends												
Phenylplastics	1 550	1 500	1 500	2 420	1 400	1 400	120	130	130	20	30	30
"Noril" blend	-	1	2	4	6	8	-	-	-	-	-	-

BULGARIA (continued)

Production, import and export of engineering plastics

(in tonnes)

Plastic	Production			Imports			Exports		
	1985	1988	1989	1985	1988	1989	1985	1988	1989
Polycarbonate	-	-	-	500	600	650	-	-	-
Polyformaldehyde	-	-	-	150	250	310	-	-	-
Polyamide and polyimide	3 000	3 100	4 200	750	800	800	-	100	2
Polybutyleneterephthalate	-	-	-	3	10	15	-	-	-
Polyethyleneterephthalate	-	15	15	10	50	90	-	-	-
Polychlortrifluoroethylene	-	-	-	2	10	10	-	-	-
Polytetrafluoroethylene	-	-	-	85	110	110	-	-	-
Polysulphone	-	-	-	2	11	12	-	-	-
Polymethylmethacrylate	-	100	100	1 600	1 850	1 900	-	-	-
Polyurethane*	8 000	10 500	9 500	180	250	280	-	-	-
Acrylonitrile-butadiene-styrene	-	-	-	450	520	280	-	-	-
Glass-fibre-reinforced	3 100	2 900	3 000	-	-	-	-	-	-
Other plastic									
Pheno-Amino-plastic	5 300	5 700	5 500	1 100	1 000	900	-	-	-
Total engineering plastics	19 400	22 317	22 319	4 837	5 469	5 647	-	100	210
Total of all plastics	387 600	327 800	320 600	-	-	-	-	-	-
Percentage of engineering plastics	5.00	6.80	6.96						

* Produced from imported raw material. About 75%-flexible-block-foampolyurethane.

CYPRUS

Production, import and export of engineering plastics

(in tonnes)

Plastic	Production			Imports			Exports		
	1985	1988	1989	1985	1988	1989	1985	1988	1989
Polycarbonate				n.a.	38	47			
Polyamide and polyimide				n.a.	23	37			
Polyethylene-terephthalate				n.a.	20	296			
Polyetherimide				n.a.	685	437			
Polyacetal				n.a.	2	7			
Polyphenylene-sulphide				n.a.	43	44			
Polysulphone				n.a.	4	14			
Polymethyl-				n.a.	9 980	11 612			
methacrylate				n.a.	2 200	2 093			

PORTUGAL

Production, import and export of engineering plastics

(in tonnes)

Plastic	Production			Imports			Exports		
	1985	1988	1989	1985	1988*	1989*	1985	1988*	1989*
Polycarbonate	–	–	–	617	1 067	1 045	1	1	1
Polyformaldehyde	–	–	–	...	...	...	...	...	...
Polyamide and polyimide	–	–	–	3 934	3 580	3 812	20	9	84
Polybutylene-terephthalate	–	–	–	46	124	232	–	–	6
Polyethylene-terephthalate	–	464	811	1 956	1 827	1 606	780	513	176
Polyetherimide	–	–	–		...	...	...	...	...
Polyacetal	–	–	–	24	547	577	–	10	1
Polytetra-fluoroethylene	–	–	–	...	2	4	...	–	–
Polyvinyl-fluoride	–	–	–	...	–	7	...	–	–
Polyvinylidene-fluoride	–	–	–						
Polymethyl-methacrylate	–	–	–	439	846	1 209	4	2	107
Polyurethane	–	–	–	1 897	1 822	2 204	13	18	27
Acrylonitrile-butadiene-styrene	–	–	–	1 121	3 682	3 224	5	1	–
Total engineering plastics	–	464	811	...	...	...	...	...	...
Total of all plastics**	(a)	(b)	(c)	(d)	(e)	(f)	(g)	(h)	(i)
Percentage of engineering plastics									

(a) 363 574; (b) 521 520; (c) 486 200; (d) 114 611; (e) 144 899; (f) 184 689; (g) 84 198; (h) 150 033; (i) 130 043.

* Preliminary figures.
** In primary forms.

SWEDEN

Production, import and export of engineering plastics

(in tonnes)

Plastic	Production			Imports			Exports		
	1985	1988	1989	1985	1988	1989	1985	1988	1989
Polycarbonate	396	635		3 084	4 964	4 865	354	748	1 416
Polyformaldehyde	-	-		2 221	3 132	3 141	3	75	207
Polyamide and polyimide	6 743	5 707		8 564	9 922	9 486	3 474	1 692	4 042
Polybutylene terephthalate	27 598	162		8 012	4 793	7 425	11 616	1 050	667
Polyethylene-terephthalate	-	-		2 097	7 512	3 884	19	86	47
Polyetherimide	5 329	5 007		11 016	4 535	4 766	2 793	603	782
Polyacetal	-	-		2 221	-	-	-	-	-
Polychlortrifluoroethylene	-	-		614	465	513	138	42	30
Polytetrafluoroethylene	-	-		-	182	204	-	13	22
Polyphenylene-oxide	2 703	2 072		18 630	19 684	18 563	14 965	1 717	1 723
Polyester-etherketone	-	162		-	-	-	-	-	-
Polymethylmethacrylate	13 283	-		26 844	2 531	2 473	11 639	357	94
Polyurethane	12 488	12 400		5 674	5 447	5 678	2 048	623	749
Acrylonitrile-butadiene-styrene	28 391	2 475		25 905	19 169	17 876	572	2 085	1 969
Total engineering plastics	96 931	28 620		114 882	85 216	78 874	47 621	9 091	11 748
Total of all plastics	628 000	842 000		618 710	614 000	637 000	418 511	455 000	457 000
Percentage of engineering plastics	15%	3%		19%	14%	12%	11%	2%	3%

UNION OF SOVIET SOCIALIST REPUBLICS

Production, import and export of engineering plastics

(in tonnes)

Plastic	Production			Imports			Exports		
	1985	1988	1989	1985	1988	1989	1985	1988	1989
Polycarbonate	2 840	2 569	2 907	-	-	-	-	-	-
Polyformaldehyde	3 338	3 280	3 167	371	945	1 303	-	-	-
Polyamide and polyimide	38 007	51 022	55 208	2 652	4 782	5 800	27	-	257
Polybutyleneterephthalate	161	292	335	72	263	101	-	50	-
Polysulphone	45	45	107	50	120	280	-	-	-

UNION OF SOVIET SOCIALIST REPUBLICS (continued)

Consumption of engineering plastics in various sectors of national economies (USSR)

(in tonnes)

Plastics	Total consumption			of which									Other		
				Engineering			Electronics			Automotive industry					
	1985	1988	1989	1985	1988	1989	1985	1988	1989	1985	1988	1989	1985	1988	1989
Total engineering plastics	51 327	70 383	78 784	31 881	46 339	49 384	3 535	4 632	5 947	4 365	5 795	7 666	11 546	13 617	15 780
of which:															
Polycarbonate	5 931	8 877	11 647	2 981	4 700	5 800	1 100	1 500	2 000	1 700	2 500	3 500	150	177	347
Polyacetal	3 709	4 225	4 470	2 000	2 135	2 180	335	400	500	937	1 250	1 350	437	440	440
Polyamide	40 659	55 804	61 008	26 500	38 904	40 908	2 000	2 500	3 000	1 200	1 400	2 100	10 959	13 000	15 000
Polybutyleneterephthalate	233	555	436	-	200	100	5	10	20	228	345	316	-	-	-
Polyethyleneterephthalate	700	700	796	400	400	396	-	-	-	300	300	400	-	-	-
Polysulphone	95	222	427	-	-	-	95	222	427	-	-	-	-	-	-
Total composites of which:	8 489	12 431	12 636	3 100	4 700	4 700	1 161	1 486	1 720	3 628	5 445	5 416	600	800	800
Polycarbonate	356	276	400	-	-	-	356	276	400	-	-	-	-	-	-
Polyamides	7 900	11 600	11 800	3 100	4 500	4 600	800	1 200	1 300	3 400	5 100	5 100	600	800	800
Polybutyleneterephthalate	233	555	436	-	200	100	5	10	20	228	345	316	-	-	-

UNITED KINGDOM

Production, import and export of engineering plastics

(in tonnes)

Plastic	Production			Imports			Exports		
	1985	1988	1989	1985	1988	1989	1985	1988	1989
Polycarbonate	0	0	0	n/a	n/a	16 400	n/a	n/a	1 100
Polyamide and polyimide	49 900	29 000	27 700	27 700	41 900	46 900	55 600	42 900	44 600
Polybutyleneterephthalate)) Polyethyleneterephthalate)	n/a	69 200	67 100	n/a	62 200	66 000	n/a	52 400	53 100
Polyacetal	0	0	0	14 100	16 400	16 200	1 300	2 700	1 600
Polytetrafluoroethylene	n/a	1 800	2 000	n/a	13 500	14 200	n/a	12 300	14 200
Polyurethane	82 400	112 400	130 700	24 800	17 300	19 800	13 200	14 700	18 500
Acrylonitrile-butadiene-styrene	63 800	104 800	115 000	33 200	51 500	53 100	42 000	87 300	94 600
Other: Acrylics	69 600	n/a	n/a	42 800	74 000	81 700	84 400	n/a	n/a
Total engineering plastics	265 700	317 200	342 500	142 600	276 800	314 300	196 500	212 300	227 700
Total of all plastics	1 958 600	1 911 200	2 032 200	1 424 200	2 191 000	2 319 800	779 800	851 500	907 000
Percentage of engineering plastics	13.6	16.6	14.9	10.0	12.6	13.5	25.2	24.9	25.1

- - - - -